Problems of Number Theory in Mathematical Competitions

Mathematical Olympiad Series

ISSN: 1793-8570

Series Editors: Lee Peng Yee *(Nanyang Technological University, Singapore)*
Xiong Bin *(East China Normal University, China)*

Published

Vol. 1   A First Step to Mathematical Olympiad Problems
by Derek Holton (University of Otago, New Zealand)

Vol. 2   Problems of Number Theory in Mathematical Competitions
by Yu Hong-Bing (Suzhou University, China)
translated by Lin Lei (East China Normal University, China)

Vol. 3   Graph Theory
by Xiong Bin (East China Normal University, China) &
Zheng Zhongyi (High School Attached to Fudan University, China)
translated by Liu Ruifang, Zhai Mingqing & Lin Yuanqing
(East China Normal University, China)

Vol. 4   Combinatorial Problems in Mathematical Competitions
by Yao Zhang (Hunan Normal University, P. R. China)

Vol. 5   Selected Problems of the Vietnamese Olympiad (1962–2009)
by Le Hai Chau (Ministry of Education and Training, Vietnam)
& Le Hai Khoi (Nanyang Technology University, Singapore)

Vol. 6   Lecture Notes on Mathematical Olympiad Courses:
For Junior Section (In 2 Volumes)
by Xu Jiagu

Vol. 7   A Second Step to Mathematical Olympiad Problems
by Derek Holton (University of Otago, New Zealand &
University of Melbourne, Australia)

Vol. 8   Lecture Notes on Mathematical Olympiad Courses:
For Senior Section (In 2 Volumes)
by Xu Jiagu

Vol. 9   Mathemaitcal Olympiad in China (2009–2010)
edited by Bin Xiong (East China Normal University, China) &
Peng Yee Lee (Nanyang Technological University, Singapore)

Yu Hong-Bing
Suzhou University, China

translated by
Lin Lei
East China Normal University, China

Vol. 2 | Mathematical
Olympiad
Series

Problems of Number Theory in Mathematical Competitions

East China Normal
University Press

World Scientific

Published by

East China Normal University Press
3663 North Zhongshan Road
Shanghai 200062
China

and

World Scientific Publishing Co. Pte. Ltd.
5 Toh Tuck Link, Singapore 596224
USA office: 27 Warren Street, Suite 401-402, Hackensack, NJ 07601
UK office: 57 Shelton Street, Covent Garden, London WC2H 9HE

Library of Congress Cataloging-in-Publication Data
Yu, Hong-Bing.
    Problems of number theory in mathematical competitions / Yu Hong-Bing ;
translated by Lin Lei.
      p. cm. -- (Mathematical Olympaid series ; v. 2)
    ISBN-13 978-981-4271-14-1 (pbk)
    ISBN-10 981-4271-14-4 (pbk)
    1. Number theory. 2. Mathematics--Competitions.
                                                        2010484270

British Library Cataloguing-in-Publication Data
A catalogue record for this book is available from the British Library.

First published 2010
Reprinted 2011, 2013

Printed in Singapore.

Introduction

Number theory is an important research field in mathematics. In mathematical competition, problems of elementary number theory occur frequently. This kind of problems uses little knowledge and has lots of variations. They are flexible and diverse.

In the book we introduce some basic concepts and methods in elementary number theory via problems in mathematics competition. We hope that readers read the book with paper and pencil, and try to solve them by themselves before they read the solutions of examples. Only in this way can they really appreciate the tricks of problem-solving.

Contents

1   Divisibility

Numbers involved in this book are integers, and letters used in this book stand for integers without further specification.

Given numbers a and b, with $b \neq 0$, if there is an integer c, such that $a = bc$, then we say b *divides* a, and write $b \mid a$. In this case we also say b is a *factor* of a, or a is a *multiple* of b. We use the notation $b \nmid a$ when b does not divide a (i.e., no such c exists).

Several simple properties of divisibility could be obtained by the definition of divisibility (proofs of the properties are left to readers).

(1) If $b \mid c$, and $c \mid a$, then $b \mid a$, that is, divisibility is transitive.

(2) If $b \mid a$, and $b \mid c$, then $b \mid (a \pm c)$, that is, the set of multiples of an integer is closed under addition and subtraction operations.

By using this property repeatedly, we have, if $b \mid a$ and $b \mid c$, then $b \mid (au + cv)$, for any integers u and v. In general, if $a_1, a_2, \ldots, a_n$ are multiples of b, then $b \mid (a_1 + a_2 + \cdots + a_n)$.

(3) If $b \mid a$, then $a = 0$ or $|a| \geqslant |b|$. Thus, if $b \mid a$ and $a \mid b$, then $|a| = |b|$.

Clearly, for any two integers a and b, a is not always divisible by b. But we have the following result, which is called the division algorithm. It is the most important result in elementary number theory.

(4) (The division algorithm) Let a and b be integers, and $b > 0$. Then there is a unique pair of integers q and r, such that

$$a = bq + r \text{ and } 0 \leqslant r < b.$$

The integer q is called the (incomplete) *quotient* when a is divided by b, r called the *remainder*. Note that the values of r has b kinds of possibilities: $0, 1, \ldots, b-1$. If $r = 0$, then a is divisible by b.

It is easy to see that the quotient q in the division algorithm is in fact $\left[\dfrac{a}{b}\right]$ (the greatest integer not exceeding $\dfrac{a}{b}$), and the heart of the division algorithm is the inequality about the remainder r: $0 \leqslant r < b$. We will go back to this point later on.

The basic method of proving $b \mid a$ is to factorize a into the product of b and another integer. Usually, in some basic problems this kind of factorization can be obtained by taking some special value in algebraic factorization equations. The following two factorization formulae are very useful in proving this kind of problems.

(5) if n is a positive integer, then

$$x^n - y^n = (x-y)(x^{n-1} + x^{n-2}y + \cdots + xy^{n-2} + y^{n-1}).$$

(6) If n is a positive odd number, then

$$x^n + y^n = (x+y)(x^{n-1} - x^{n-2}y + \cdots - xy^{n-2} + y^{n-1}).$$

Example 1   Prove that $\underbrace{10\cdots01}_{200}$ is divisible by 1001.

Proof   By factorization formula (6), we have

$$\underbrace{10\cdots01}_{200} = 10^{201} + 1 = (10^3)^{67} + 1$$

$$= (10^3 + 1)[(10^3)^{66} - (10^3)^{65} + \cdots - 10^3 + 1].$$

Therefore, $10^3 + 1 (= 1001)$ divides $\underbrace{10\cdots01}_{200}$.

Example 2   Let $m > n \geqslant 0$, show that $(2^{2^n} + 1) \mid (2^{2^m} - 1)$.

Proof   Take $x = 2^{2^{n+1}}$, $y = 1$ in factorization (5), and substitute n by 2^{m-n-1}, we get

$$2^{2^m} - 1 = (2^{2^{n+1}} - 1)[(2^{2^{n+1}})^{2^{m-n-1}-1} + \cdots + 2^{2^{n+1}} + 1].$$

Thus,

$$(2^{2^{n+1}} - 1) \mid (2^{2^m} - 1).$$

But

$$2^{2^{n+1}} - 1 = (2^{2^n} - 1)(2^{2^n} + 1).$$

Hence,

$$(2^{2^n} + 1) \mid (2^{2^{n+1}} - 1).$$

Further, by property (1) we have $(2^{2^n} + 1) \mid (2^{2m} - 1)$.

Remark Sometimes it is difficult to prove $b \mid a$ directly when dealing with divisibility problems. Therefore, we can attempt to choose an "intermediate number" c and prove $b \mid c$ and $c \mid a$ first, then use the property (1) of divisibility to deduce the conclusion.

Example 3 For a positive integer n, write $S(n)$ to denote the sum of digits appearing in the expression of n in base 10. Show that $9 \mid n$ if and only if $9 \mid S(n)$.

Proof Write $n = a_k \times 10^k + \cdots + a_1 \times 10 + a_0$ (where $0 \leqslant a_i \leqslant 9$, and $a_k \neq 0$), then $S(n) = a_0 + a_1 + \cdots + a_k$. We have

$$n - S(n) = a_k(10^k - 1) + \cdots + a_1(10 - 1). \qquad (1.1)$$

For $1 \leqslant i \leqslant k$, from factorization (5) we get $9 \mid (10^i - 1)$. So every term of the k terms in the right-hand side of equation (1.1) is a multiple of 9, thus property (2) implies that their sum is also a multiple of 9, that is, $9 \mid (n - S(n))$. Hence, the result can be obtained easily.

Remark 1 The divisibility property (2) provides an elementary method to prove $b \mid (a_1 + a_2 + \cdots + a_n)$. We can try to prove a stronger statement (which is usually easier to prove): b divides every a_i ($i = 1$, $2, \ldots, n$).

Of course this stronger statement does not always hold true. But even if it does not hold true, the above method is also useful. We can rewrite the sum $a_1 + a_2 + \cdots + a_n$ into $c_1 + c_2 + \cdots + c_k$ by regrouping the numbers, then we need to prove $b \mid c_i$ ($i = 1, 2, \ldots, k$). Readers will find out that in order to solve some special problems, sometimes we can express a as a sum of certain numbers, and then apply the above method to prove it.

Remark 2 From the proof of Example 3 we actually obtain a stronger conclusion, that is, the difference between n and $S(n)$ is always a multiple of 9. So n and $S(n)$ have the same remainder when divided by 9 (so we say n is congruent to $S(n)$ mod (9). Please refer to Chapter 6 for details).

Remark 3   In some cases from the properties of digits base 10 of a positive integer we can judge whether or not this integer is divisible by another integer. This kind of results sometimes are called "the digit character of divisibility". The digit characters of an integer divisible by 2, 5 and 10 are well-known. In Example 3 we present the digit character of an integer divisible by 9. For this result there are many applications. In addition, in Exercise 1.3 the digit character of an integer divisible by 11 is given. This result is useful too.

Example 4   Let $k \geqslant 1$ be odd. Prove that for any positive integer n, $1^k + 2^k + \cdots + n^k$ is not divisible by $n + 2$.

Proof   When $n = 1$ the statement is obviously true. For $n \geqslant 2$, denote the sum by A, then

$$2A = 2 + (2^k + n^k) + (3^k + (n-1)^k) + \cdots + (n^k + 2^k).$$

Since k is a positive odd number, from formula (6) we know that for every $i \geqslant 2$, $i^k + (n+2-i)^k$ is divisible by

$$i + (n + 2 - i) = n + 2.$$

Thus $2A$ has remainder 2 when divided by $n + 2$, which implies that A is not divisible by $n + 2$ (note that $n + 2 > 2$).

Remark   In the proof we use the "pairing method" which is a common method to transform the expression of a sum.

Example 5   Let m and n be positive integers with $m > 2$. Prove that $(2^m - 1) \nmid (2^n + 1)$.

Proof   At first, when $n \leqslant m$ it is easy to prove that the result is true. In fact, when $m = n$ the result is trivial. When $n < m$ from inequalities

$$2^n + 1 \leqslant 2^{m-1} + 1 < 2^m - 1,$$

we can get the result (note that $m > 2$ and refer to the divisibility property (3)).

Secondly, we can reduce the case $n > m$ to the special situation above: by the division algorithm, $n = mq + r$, $0 \leqslant r < m$, and $q > 0$. Since

$$2^n + 1 = (2^{mq} - 1)2^r + 2^r + 1,$$

we know $(2^m - 1) \mid (2^{mq} - 1)$ by factorization (5). But $0 \leqslant r < m$, from the discussion above we get $(2^m - 1) \nmid (2^n + 1)$ (note that when $r = 0$ the result is trivial). Hence, when $n > m$ we also have $(2^m - 1) \nmid (2^n + 1)$. The proof is complete.

Exercises

1.1 Let n and k be positive integers, then among numbers $1, 2, \ldots, n$ there are exactly $\left[\dfrac{n}{k}\right]$ numbers which are divisible by k.

1.2 11 girls and n boys go to pick mushrooms. All the children pick $n^2 + 9n - 2$ mushrooms in total, and every child picks the equal number of mushrooms. Are there more girls or more boys among these children?

1.3 Let n be a positive number, and n can be expressed as $\overline{a_k \cdots a_1 a_0}$ (where $0 \leqslant a_i \leqslant 9$, $a_k \neq 0$). Set

$$T(n) = a_0 - a_1 + \cdots + (-1)^k a_k$$

(the alternating sum of the digits of n beginning with the units digit of n). Show that 11 divides $n - T(n)$, which implies that the digit character of an integer divisible by 11 is: 11 divides n if and only if 11 divides $T(n)$.

1.4 Suppose that there are n integers which have the following property: the difference between the product of any $n - 1$ integers and the remaining one is divisible by n. Prove that the sum of the square of these n numbers is also divisible by n.

1.5 Let a, b, c, d be integers with $ad - bc > 1$. Prove that there is at least one among a, b, c, d which is not divisible by $ad - bc$.

2 Greatest Common Divisors and Least Common Multiples

Greatest common divisor is an important concept.

Let a, b be not both zero. An integer which can divide both a and b (for instance, ± 1) is called a common divisor of a and b. Since a, b are not both zero, property (3) of Chapter 1 implies that there exist only a finite number of common divisors of a, b. The greatest one among them is called the greatest common divisor of a, b, and denoted by $\gcd(a, b)$. Clearly, the greatest common divisor is a positive integer.

If $\gcd(a, b) = 1$ (that is, common divisors of a, b are only ± 1), we say that a and b are relatively prime. Readers can find in the sequel that this case is particularly important.

For more than two integers (not all zero) a, b, $\ldots$, c, we can define their greatest common divisor $\gcd(a, b, \ldots, c)$ similarly. If $\gcd(a, b, \ldots, c) = 1$, then a, b, $\ldots$, c are called relatively prime. Attention: if this is the case, usually we cannot deduce that a, b, $\ldots$, c are relatively prime pairwise (that is, any two of them are relatively prime). On the other hand, if a, b, $\ldots$, c are relatively prime pairwise, then clearly, $\gcd(a, b, \ldots, c) = 1$.

By definition of greatest common divisor, we have some simple properties as follows.

If we change signs of a and b, the value of $\gcd(a, b)$ does not change. That is, $\gcd(\pm a, \pm b) = \gcd(a, b)$.

The expression $\gcd(a, b)$ is symmetric for a and b, that is, $\gcd(a, b) = \gcd(b, a)$.

$\gcd(a, b)$ as a function of variable b is periodic, a is its period, that is, $\gcd(a, b + ax) = \gcd(a, b)$, for any integer x.

The result (1) below is fundamental for getting more properties about greatest common divisors.

(1) Let a and b be integers not both zero. Then there exist integers x and y such that

$$ax + by = \gcd(a, b).$$

We remark that if $x = x_0$ and $y = y_0$ are a pair of integers satisfying the above equation, then the equation

$$a(x_0 + bu) + b(y_0 - au) = \gcd(a, b) \quad (u \text{ is any integer})$$

implies that such pairs of integers x and y are infinitely many. Further, if $ab > 0$, we can choose one pair such that x is positive (negative) and y is negative (positive).

From result (1) we can easily get the following result (2).

(2) A necessary and sufficient condition that two integers a and b are relatively prime is: there exist integers x and y, such that

$$ax + by = 1.$$

Usually, this is called Bézout's identity.

In fact, the necessary condition is a special case of (1). On the other hand, suppose there exist integers x and y such that the above equation holds. Let $\gcd(a, b) = d$. Then $d \mid a$ and $d \mid b$, so $d \mid ax$ and $d \mid by$. Thus $d \mid (ax + by)$, it means $d \mid 1$. Therefore $d = 1$.

By (1) and (2) we obtain the following basic results easily.

(3) If $m \mid a$ and $m \mid b$, then $m \mid \gcd(a, b)$, it means that every common divisor of a and b is a factor of their greatest common divisor.

(4) If $m > 0$, then $\gcd(ma, mb) = m \gcd(a, b)$.

(5) If $\gcd(a, b) = d$, then $\left(\dfrac{a}{d}, \dfrac{b}{d}\right) = 1$. Hence, from any two integers which are not relatively prime we can get a pair of relatively prime integers naturally.

(6) If $\gcd(a, m) = 1$ and $\gcd(b, m) = 1$, then $\gcd(ab, m) = 1$. It means that the set of integers which is relatively prime with a fixed integer is closed under multiplication. From this fact we know that if

$\gcd(a, b) = 1$, then for any $k > 0$, $\gcd(a^k, b) = 1$. Thus, $\gcd(a^k, b^l) = 1$, for any $l > 0$.

(7) Suppose $b \mid ac$. If $\gcd(b, c) = 1$, then $b \mid a$.

(8) Suppose that the product of positive integers a and b is a k-th power of an integer $(k \geq 2)$. If $\gcd(a, b) = 1$, then a and b are all k-th power of integers. In general, suppose that the product of positive integers $a, b, \ldots, c$ is a k-th power of an integer, and if $a, b, \ldots, c$ are relatively prime pairwise, then $a, b, \ldots, c$ are all k-th power of integers.

Properties (6), (7) and (8) show the importance of relatively prime. They have many applications.

Now we discuss least common multiples briefly.

Let a and b be integers not both zero, an integer which is a multiple of both a and b is called a common multiple of a, b. Clearly, there are infinitely many common multiples of a and b, the least positive number among them is called the least common multiple of a and b, denoted by $[a, b]$. For more than two non-zero integers $a, b, \ldots, c$, we can define their least common multiple $[a, b, \ldots, c]$ similarly.

Here are some main properties of least common multiples.

(9) Any common multiple of a and b is a multiple of $[a, b]$.

(10) For any two integers a and b, their greatest common divisor and their least common multiple satisfy the following identity

$$\gcd(a, b)[a, b] = |ab|.$$

However note that for the case of more than two integers, we cannot get a similar result (readers can give some examples by themselves). On the other hand, we have the following conclusion.

(11) If $a, b, \ldots, c$ are relatively prime pairwise, then

$$[a, b, \ldots, c] = |ab\cdots c|.$$

From this and (9) we know that if $a \mid d$, $b \mid d$, $\ldots$, $c \mid d$, and $a, b, \ldots, c$ are relatively prime pairwise, then $ab\cdots c \mid d$.

The concept of relatively prime is very important in number theory, and is the key or basis of many problems.

In mathematical competition, we have to prove that two integers are relatively prime (or determine their greatest common divisor) in some problems. From examples below we show some fundamental methods in dealing with such problems.

Example 1   For any integer n, prove that the fraction $\dfrac{21n+4}{14n+3}$ is irreducible.

Proof   We have to prove that $21n+4$ and $14n+3$ are relatively prime. It is clear that these two integers satisfy the following identity (i.e., Bézout's identity)

$$3(14n+3) - 2(21n+4) = 1.$$

Hence, we obtain the conclusion required.

In general, it is not easy to get Bézout's identity for relatively prime integers a and b, therefore we often use the following alternative way: to create an auxiliary equation which is similar to Bézout's identity

$$ax + by = r,$$

where r is an appropriate integer. If $\gcd(a, b) = d$, then from the above equation we get $d \mid r$. The so-called appropriate r means: from $d \mid r$ we can derive $d = 1$ through further proof, or alternatively the number of divisors of r is relatively small, and we can get the result by exclusion method.

In addition, the auxiliary equation above is equivalent to $a \mid (by - r)$ or $b \mid (ax - r)$, sometimes, these can be derived more easily by divisibility.

Example 2   Let n be a positive integer. Prove that

$$\gcd(n! + 1, (n+1)! + 1) = 1.$$

Proof   We have the following equation

$$(n! + 1)(n+1) - ((n+1)! + 1) = n. \tag{2.1}$$

Set

$$d = \gcd(n! + 1, (n+1)! + 1),$$

then from (2.1) we have $d \mid n$.

Further, $d \mid n$ implies that $d \mid n!$, and combine with $d \mid (n! + 1)$ we know $d \mid 1$, thus $d = 1$.

Example 3 Set $F_k = 2^{2^k} + 1$, $k \geq 0$. Prove that if $m \neq n$, then $\gcd(F_m, F_n) = 1$.

Proof By symmetry, we can assume that $m > n$. The key of the proof is to use the fact $F_n \mid (F_m - 2)$ (cf. Example 2 of Chapter 1), i.e., there is an integer x such that

$$F_m + xF_n = 2.$$

Put $d = \gcd(F_m, F_n)$, from the above equation we get $d \mid 2$, so $d = 1$ or 2. But F_n is clearly odd, thus $d = 1$.

Remark $F_k (k \geq 0)$ is called Fermat number. Example 3 shows that Fermat numbers are relatively prime pairwise. This is an interesting elementary property for Fermat numbers.

The conclusion in Example 4 below has many uses, it is worth to remember.

Example 4 Let $a > 1$, m, $n > 0$, prove that

$$\gcd(a^m - 1, a^n - 1) = a^{\gcd(m, n)} - 1.$$

Proof Put $D = \gcd(a^m - 1, a^n - 1)$. We can get the equation $D = a^{\gcd(m, n)} - 1$ by proving $(a^{\gcd(m, n)} - 1) \mid D$ and $D \mid (a^{\gcd(m, n)} - 1)$. This is a common manner of proving that two numbers are equal in number theory.

Since $\gcd(m, n) \mid m$ and $\gcd(m, n) \mid n$, by decomposition formula (5) in Chapter 1 we have

$$(a^{\gcd(m, n)} - 1) \mid (a^m - 1)$$

and

$$(a^{\gcd(m, n)} - 1) \mid (a^n - 1).$$

Thus, property (3) implies that $a^{\gcd(m, n)} - 1$ divides $\gcd(a^m - 1, a^n - 1)$, that is, $(a^{\gcd(m,\, n)} - 1) \mid D$.

In order to prove $D \mid (a^{\gcd(m,\, n)} - 1)$, we set $d = \gcd(m, n)$. Since $m, n > 0$, we can choose $u, v > 0$ such that (cf. the Remark in property (1))

$$mu - nv = d. \tag{2.2}$$

Since $D \mid (a^m - 1)$, $D \mid (a^{mu} - 1)$. In the same way, $D \mid (a^{nv} - 1)$. Thus, $D \mid (a^{mu} - a^{nv})$. Now due to (2.2) we have

$$D \mid a^{nv}(a^d - 1). \tag{2.3}$$

On the other hand, since $a > 1$ and $D \mid (a^m - 1)$, so $\gcd(D, a) = 1$, and $\gcd(D, a^{nv}) = 1$. Hence, by (2.3) and property (7) we have $D \mid (a^d - 1)$. Therefore, $D \mid (a^{\gcd(m,\, n)} - 1)$.

Combining with the two aspects of the results proved we have $D = a^{\gcd(m,\, n)} - 1$.

Example 5 Let $m, n > 0$, $mn \mid (m^2 + n^2)$, then $m = n$.

Proof Put $\gcd(m, n) = d$, then $m = m_1 d$, $n = n_1 d$, where $\gcd(m_1, n_1) = 1$. Thus, the given condition is reduced to $m_1 n_1 \mid (m_1^2 + n_1^2)$, which implies that $m_1 \mid (m_1^2 + n_1^2)$. Hence, $m_1 \mid n_1^2$. But $\gcd(m_1, n_1) = 1$, thus $\gcd(m_1, n_1^2) = 1$. Combining with $m_1 \mid n_1^2$, we have $m_1 = 1$. Similarly, $n_1 = 1$. Therefore, $m = n$.

Remark 1 For two given integers not all zero, we often make use of their greatest common divisor, and apply property (5) to get two relatively prime integers. Hence make further deduction by using the fact of being relatively prime (cf. properties (6) and (7)). For this example, mn is quadratic, and $m^2 + n^2$ is a quadratic homogeneous expression. Essentially, the effect of these procedures is to reduce the problem to a special case when m and n are relatively prime.

Remark 2 In some problems, the given condition (or the result just obtained) $c \mid a$ is not applicable. We can try to choose a suitable divisor of c, and from $c \mid a$ we get $b \mid a$ (a weaker conclusion). Then we expect that the latter may provide suitable information for further proof. In Example 5, from the fact $m_1 n_1 \mid (m_1^2 + n_1^2)$ we have $m_1 \mid n_1^2$, which implies $m_1 = 1$.

Example 6   Suppose that the greatest common divisor of the positive integers a, b and c is 1, and

$$\frac{ab}{a-b} = c.$$

Prove that $a - b$ is a perfect square.

Proof   Set $\gcd(a, b) = d$, then $a = da_1$, $b = db_1$, where $\gcd(a_1, b_1) = 1$. Since $\gcd(a, b, c) = 1$, $\gcd(d, c) = 1$.

Now, the identity in the problem is equivalent to

$$da_1b_1 = ca_1 - cb_1, \qquad (2.4)$$

hence, a_1 divides cb_1. Since $\gcd(a_1, b_1) = 1$, we have $a_1 \mid c$. Similarly, $b_1 \mid c$. Again from $\gcd(a_1, b_1) = 1$ we get $a_1b_1 \mid c$ (cf. properties (9) and (10)).

Let $c = a_1b_1k$, where k is a positive integer. On one hand, it is clear that k divides c. On the other hand, combining with (2.4), we have $d = k(a_1 - b_1)$, so $k \mid d$. Thus, $k \mid \gcd(c, d)$ (cf. property (3)). But $\gcd(c, d) = 1$, so $k = 1$.

Therefore $d = a_1 - b_1$, we have $a - b = d(a_1 - b_1) = d^2$. It shows that $a - b$ is a perfect square.

Remark   By using primes, we have a more direct proof for this problem (cf. Exercise 3.5).

Example 7   Let k be a positive odd number. Prove that $1 + 2 + \cdots + n$ divides $1^k + 2^k + \cdots + n^k$.

Proof   In view of

$$1 + 2 + \cdots + n = \frac{n(n+1)}{2},$$

this problem is equivalent to proving that $n(n + 1)$ divides $2(1^k + 2^k + \cdots + n^k)$. Since n and $n + 1$ are relatively prime, this is also equivalent to proving that

$$n \mid 2(1^k + 2^k + \cdots + n^k)$$

and

$$(n+1) \mid 2(1^k + 2^k + \cdots + n^k).$$

In fact, since k is odd, by factorization (6) in Chapter 1 we have

$$2(1^k + 2^k + \cdots + n^k)$$
$$= [1^k + (n-1)^k] + [2^k + (n-2)^k] + \cdots + [(n-1)^k + 1^k] + 2n^k$$

is a multiple of n. Similarly,

$$2(1^k + 2^k + \cdots + n^k) = [1^k + n^k] + [2^k + (n-1)^k] + \cdots + [n^k + 1^k],$$

which is a multiple of $n+1$.

Remark   When you deal with a problem about divisibility, sometimes it is not easy to prove $b \mid a$ directly. If b has a factorization $b = b_1 b_2$, where $\gcd(b_1, b_2) = 1$, then we can divide the statement $b \mid a$ into two equivalent statements $b_1 \mid a$ and $b_2 \mid a$. The later is easier to prove. In Example 7 we have used this method. In Example 6 we do the same when we prove $a_1 b_1 \mid c$.

More generally, in order to prove $b \mid a$, we can factor b into a product of some pairwise relatively prime integers $b_1, b_2, \ldots, b_n$, and the statement is equivalent to $b_i \mid a$, for $i = 1, 2, \ldots, n$ (cf. property (11), and compare the idea in Remark 1 following Example 3 of Chapter 1). For standard application of this method, refer to Example 5 of Chapter 3.

Exercises

2.1   Let n be an integer. Prove that $\gcd(12n + 5, 9n + 4) = 1$.

2.2   Let m and n be positive integers with m odd. Prove that $\gcd(2^m - 1, 2^n + 1) = 1$.

2.3   Let $\gcd(a, b) = 1$. Prove that $\gcd(a^2 + b^2, ab) = 1$.

2.4   If the k-th power of a rational number is an integer ($k \geqslant 1$), then this rational number must be an integer. More generally, show that rational roots of a polynomial of integer coefficients with leading coefficient ± 1 are integers.

2.5   Let m, n and k be positive integers with $[m+k, m] = [n+k, n]$. Prove that $m = n$.

3 Prime Numbers and Unique Factorization Theorem

For every integer n great than 1 there are two different positive factors: 1 and n. If n has only these two positive factors (n is said to have no proper factors), we call n a prime number (or a prime). If n has proper factors, that is, n can be expressed as $a \cdot b$ (where a and b are two integers all great than 1), then n is said to be composite. Thus, positive integers can be divided into three classes: the first class which contains only number 1, the second one is the class of primes, and the third one is the class of composites.

Primes play a very important role in positive integers. Usually, we denote a prime by letter p. By definition, it is easy to get the following basic conclusions:

(1) For every integer great than 1 there exists at least one prime factor.

This is true because for every integer great than 1 there are positive divisors great than 1, the least one among them must not have proper factor, therefore it is prime.

(2) Let p be a prime, and n any integer. Then p divides n, or p and n are relatively prime.

In fact, the great common divisor $\gcd(p, n)$ of p and n divides p, so the definition of primes gives $\gcd(p, n) = 1$, or $\gcd(p, n) = p$, i.e., p and n are relatively prime, or $p \mid n$.

The most powerful property of primes is the following.

(3) Let p be a prime, a and b integers. If $p \mid ab$, then p divides a or b.

In fact, if p does not divide a and b, then by property (2), p and a are relatively prime, p and b are relatively prime too. Thus, p and *ab* are relatively prime (see property (6) in Chapter 2), this contradicts the given condition $p \mid ab$.

By (3), if a prime p divides a^n for some integer $n \geq 1$, then $p \mid a$.

One of the oldest results was proved by Euclid (in Book IX of his *Elements*):

(4) There are infinitely many primes.

We prove it by contradiction. Suppose that there are only finitely many primes, say p_1, p_2, ..., p_k. Consider the number $N = p_1 p_2 \cdots p_k + 1$. Since $N > 1$, N has a prime factor p. But p_1, p_2, ..., p_k are only primes, so p is one of $p_i (1 \leq i \leq k)$, thus, p divides $N - p_1 p_2 \cdots p_k$, it follows that p divides 1, which is impossible. So there must be infinitely many primes. (Note that $p_1 p_2 \cdots p_k + 1$ may not be a prime.)

The statement (4) can also be deduced from Example 3 in Chapter 2: Let $F_k = 2^{2^k} + 1$ ($k \geq 0$), then $F_k > 1$, so F_k has prime factors. Since we have proved that terms of the infinite sequence $\{F_k\}$ ($k \geq 0$) are pairwise relatively prime, prime factors in every F_k are different from those in other terms. Hence there must be infinitely many primes.

Now we turn to consider the most fundamental result in elementary number theory, that is, the unique factorization theorem of positive integers, or the fundamental Theorem of Arithmetic. It explains why prime numbers are so important in the set of positive integers.

(5) (The unique factorization theorem) Every positive integer great than 1 can be factorized into a product of finitely many primes, and this factorization is unique, apart from permutations of the factors.

In other words, let $n > 1$, then n can be expressed as $n = p_1 p_2 \cdots p_k$, where $p_i (1 \leq i \leq k)$ are all primes. Further, if there are another such factorization

$$n = p_1 p_2 \cdots p_k = q_1 q_2 \cdots q_l,$$

then we have $k = l$, and p_1, p_2, ..., p_k is a permutation of q_1,

$q_2, \ldots, q_l.$

If we put the same prime factors in prime factorization of n together, we know every positive integer $n > 1$ can be uniquely expressed as

$$n = p_1^{\alpha_1} p_2^{\alpha_2} \cdots p_k^{\alpha_k},$$

where $p_1, p_2, \ldots, p_k$ are different primes, $\alpha_1, \alpha_2, \ldots, \alpha_k$ are positive integers. It is called the standard factorization of n.

If the standard factorization of n is given, then from the unique factorization theorem we can determine all positive divisors of n:

(6) All positive divisors of n are $p_1^{\beta_1} p_2^{\beta_2} \cdots p_k^{\beta_k}$, where β_i are any non-negative integers with $0 \leqslant \beta_i \leqslant \alpha_i$, for $i = 1, \ldots, k$.

By (6), it is easy to see if $\tau(n)$ denotes the number of positive divisors of n, $\sigma(n)$ the sum of positive divisors of n, then we have

$$\tau(n) = (\alpha_1 + 1)(\alpha_2 + 1)\cdots(\alpha_k + 1),$$

and

$$\sigma(n) = \frac{p_1^{\alpha_1+1} - 1}{p_1 - 1} \cdot \frac{p_2^{\alpha_2+1} - 1}{p_2 - 1} \cdot \cdots \cdot \frac{p_k^{\alpha_k+1} - 1}{p_k - 1}.$$

Although there are infinitely many prime numbers, their distribution in natural numbers is extremely irregular (cf. Exercise 1.3). Given a big prime, determining whether it is prime, is usually extremely difficult, and giving its standard factorization is even more difficult. The following property (7) is very interesting, because the standard factorization of $n!$ is given for any $n > 1$.

(7) For any positive integer m and prime p, the symbol $p^\alpha \parallel m$ means $p^\alpha \mid m$, but $p^{\alpha+1} \nmid m$, that is, p^α is the power of p occurring in the standard factorization of m.

If $n > 1$ and p is prime, and $p^{\alpha_p} \parallel n!$, then

$$\alpha_p = \sum_{l=1}^{\infty} \left[\frac{n}{p^l}\right] \left(= \left[\frac{n}{p}\right] + \left[\frac{n}{p^2}\right] + \cdots \right),$$

where $[x]$ denotes the greatest integer $i \leqslant x$. Note that when $p^l > n$,

$\left[\dfrac{n}{p^l}\right] = 0$, so in the above sum there are only finitely many non-zero terms.

Proving that some special numbers are not prime (or giving the necessary condition that it is prime) is a basic problem in elementary number theory. It often occurs in mathematical competition. A main method dealing with this kind of problems is using (various) factorizations, and finding a proper divisor of the number concerned. Let us give some examples.

Example 1   Show that among the infinite sequence 10 001, 100 010 001, ... , there is no prime number.

Proof   Let $a_n = 10\,001 \cdots 10\,001$ ($n \geq 2$, the number of "1" is n), then

$$a_n = 1 + 10^4 + 10^8 + \cdots + 10^{4(n-1)} = \frac{10^{4n} - 1}{10^4 - 1}.$$

In order to factor the number on the right side of the above equation into a product of two (greater than 1) integers, we consider two cases.

The case when n is even. Let $n = 2k$, then

$$a_{2k} = \frac{10^{8k} - 1}{10^4 - 1} = \frac{10^{8k} - 1}{10^8 - 1} \cdot \frac{10^8 - 1}{10^4 - 1}.$$

It is easy to know that $\dfrac{10^8 - 1}{10^4 - 1}$ is an integer greater than 1, and for $k \geq 2$, $\dfrac{10^{8k} - 1}{10^8 - 1}$ is also an integer greater than 1. So a_{2k} ($k = 2, 3, \ldots$) are all composite, and $a_2 = 10\,001 = 13 \times 137$ is composite.

The case when n is odd. Let $n = 2k + 1$, then

$$a_{2k+1} = \frac{10^{4(2k+1)} - 1}{10^4 - 1} = \frac{10^{2(2k+1)} - 1}{10^2 - 1} \cdot \frac{10^{2(2k+1)} + 1}{10^2 + 1}$$

which is a product of two integers both greater than 1. So a_{2k+1} are all composite. Therefore, all a_n are composite.

Remark   In the proof of Example 1, the special factorization of the number is realized by using decompositions of algebraic expressions (cf. factorizations (5) and (6)), which is the same as in the following Example 2.

Example 2   Prove that for any integer $n > 1$, $n^4 + 4^n$ is not prime.

Proof   If n is even, then $n^4 + 4^n$ is greater than 2, and divisible by 2, thus not prime. But for odd number n, it is easy to know that $n^4 + 4^n$ does not have any (greater than 1) fixed divisor. So we have to deal with it by a different way.

Let odd number $n = 2k + 1$, $k \geqslant 1$, then

$$
\begin{aligned}
n^4 + 4^n &= n^4 + 4 \cdot 4^{2k} = n^4 + 4 \cdot (2^k)^4 \\
&= n^4 + 4n^2 \cdot (2^k)^2 + 4 \cdot (2^k)^4 - 4n^2 \cdot (2^k)^2 \\
&= (n^2 + 2 \cdot 2^{2k})^2 - (2 \cdot n \cdot 2^k)^2 \\
&= (n^2 + 2^{k+1}n + 2^{2k+1})(n^2 - 2^{k+1}n + 2^{2k+1}).
\end{aligned}
$$

The first factor of the right side in the above equation is clearly not equal to 1, and the second one is $(n - 2^k)^2 + 2^{2k}$ which does not equal 1 (for $k \geqslant 1$), too. Thus $n^4 + 4^n$ is composite for all $n > 1$.

Example 2 looks nothing out of the ordinary, but when you do it by yourself it may not be so smooth. The key of this solution is, when n is odd, we can regard 4^n as a monomial $4y^4$, and by using factorization of algebraic expression

$$
x^4 + 4y^4 = (x^2 + 2y^2 + 2xy)(x^2 + 2y^2 - 2xy),
$$

we can get a suitable factorization of the number.

Example 3   Suppose that positive integers a, b, c and d satisfy $ab = cd$. Prove that $a + b + c + d$ is not prime.

Proof One   For this problem it is not suitable to perform factorization of an algebraic expression to deduce some factorization required. Our first solution is by using factorizations of numbers we can find out a proper divisor of $a + b + c + d$.

From $ab = cd$, we can let $\dfrac{a}{c} = \dfrac{d}{b} = \dfrac{m}{n}$, where m and n are relatively prime positive integers. $\dfrac{a}{c} = \dfrac{m}{n}$ means the rational number

$\frac{a}{c}$ equals irreducible fraction $\frac{m}{n}$. Hence, there is a positive integer u, such that

$$a = mu,\ c = nu. \tag{3.1}$$

Similarly, there is a positive integer v, such that

$$b = nv,\ d = mv. \tag{3.2}$$

Therefore,

$$a + b + c + d = (m + n)(u + v)$$

is a product of two integers all greater than 1, and it is not prime.

Remark   If positive integers a, b, c and d satisfy $ab = cd$, then a, b, c and d can be factored into the forms (3.1) and (3.2). This result is very useful in some problems.

Proof Two   From $ab = cd$ we have $b = \frac{cd}{a}$. Hence

$$a + b + c + d = a + \frac{cd}{a} + c + d = \frac{(a + c)(a + d)}{a}.$$

Since $a + b + c +$ is an integer, $\frac{(a + c)(a + d)}{a}$ is also an integer. If it is prime, we denote it by p. Then from

$$(a + c)(a + d) = ap \tag{3.3}$$

we know p divides $(a + c)(a + d)$, thus the prime p divides $a + c$ or $a + d$. Let us assume that $p \mid (a + c)$, then $a + c \geqslant p$. Combining with (3.3), we have $a + d \leqslant a$, but this is impossible (for $d \geqslant 1$).

The method in Proof Two is using the properties of primes (cf. (3)) and contradiction, this is quite different from methods before.

Example 4   Prove that if positive integers a and b satisfy $2a^2 + a = 3b^2 + b$, then $a - b$ and $2a + 2b + 1$ are perfect squares.

Proof   The given equation is equivalent to

$$(a - b)(2a + 2b + 1) = b^2. \tag{3.4}$$

The key of the proof is to prove that $a - b$ and $2a + 2b + 1$ are relatively prime. Let

$$d = \gcd(a - b, 2a + 2b + 1).$$

If $d > 1$, then there is a prime divisor p of d. From (3.4) we get $p \mid b^2$. But p is prime, thus $p \mid b$. Now $p \mid (2a + 2b + 1)$ implies that $p \mid 1$, which is impossible. Hence, $d = 1$, and from (3.4) we know positive integers $a - b$ and $2a + 2b + 1$ are all perfect squares (cf. (8) in Chapter 2).

Remark We need to prove that a certain positive integer is 1 in many problems concerning number theory (for example, proving that the greatest common divisor of some integers is 1). Property (1) supplies us with a description in a number theory way whether an integer equals 1. Hence, we usually assume that there is a prime divisor for a given number, and do further proof by using the "sharp" property (3), then give a contradiction. Example 4 is a such case.

Example 5 Let n, a and b be integers, and $a \neq b$. Prove that $n \left| \dfrac{a^n - b^n}{a - b} \right.$ if $n \mid (a^n - b^n)$.

Proof Let p be a prime, and $p^\alpha \parallel n$. We want to show that $p^\alpha \left| \dfrac{a^n - b^n}{a - b} \right.$, which gives the required conclusion (cf. the Remark below).

Set $t = a - b$. If $p \nmid t$, then $\gcd(p^\alpha, t) = 1$. Since $n \mid (a^n - b^n)$, we have $p^\alpha \mid (a^n - b^n)$. But $a^n - b^n = t \cdot \dfrac{a^n - b^n}{t}$, thus $p^\alpha \left| \dfrac{a^n - b^n}{t} \right.$.

If $p \mid t$, then using the binomial theorem, we get

$$\frac{a^n - b^n}{t} = \frac{(b + t)^n - b^n}{t} = \sum_{i=1}^{n} \binom{n}{i} b^{n-i} t^{i-1}. \qquad (3.5)$$

Let $p^\beta \parallel i$ $(i \geqslant 1)$, then $2^\beta \leqslant p^\beta \leqslant i$, thus $\beta \leqslant i - 1$. Therefore, the power of p containing in

$$\binom{n}{i} t^{i-1} = \frac{n}{i} \binom{n-1}{i-1} t^{i-1}$$

is at least

$$\alpha - \beta + (i - 1) \geqslant \alpha,$$

and every term on the right side of (3.5) is divisible by p^{α}. So $p^{\alpha} \left| \dfrac{a^n - b^n}{t} \right.$, that is, $p^{\alpha} \left| \dfrac{a^n - b^n}{a - b} \right.$ (cf. Remark 1 following Example 3 in Chapter 1).

Remark   In order to prove $b \mid a$, we can factor b into the standard form $b = p_1^{q_1} p_2^{q_2} \cdots p_k^{q_k}$, thus reduce the problem to proving $p_i^{q_i} \mid a$ $(i = 1, 2, \ldots, k)$ (cf. property (11) in Chapter 2). The benefit to do it this way is that we can apply the sharp property of primes. The proof of Example 5 shows it clearly.

Example 6   Let m and n be non-zero integers. Prove that $\dfrac{(2m)!(2n)!}{m!n!(m + n)!}$ is an integer.

Proof   We just need to prove that for every prime p, the power of p occurring in the standard factorization of the denominator $m!n!(m + n)!$ is not greater than that of p occurring in the standard factorization of the numerator $(2m)!(2n)!$. By formula in (7) we know this is equivalent to proving

$$\sum_{l=1}^{\infty} \left(\left[\frac{2m}{p^l} \right] + \left[\frac{2n}{p^l} \right] \right) \geqslant \sum_{l=1}^{\infty} \left(\left[\frac{m}{p^l} \right] + \left[\frac{n}{p^l} \right] + \left[\frac{m + n}{p^l} \right] \right). \qquad (3.6)$$

In fact, we can prove the following stronger result: For any real numbers x and y, we have

$$[2x] + [2y] \geqslant [x] + [y] + [x + y]. \qquad (3.7)$$

To prove (3.7), we note that for any integer k and any real number α, $[k + \alpha] = [\alpha] + k$ holds. From this, it is easy to know that if x or y changes an integer quantity, then two sides in the inequality (3.7) change the same quantity. Therefore, it is sufficient to prove the formula (3.7) for the case when $0 \leqslant x < 1$ and $0 \leqslant y < 1$. Thus it reduces to prove the inequality

$$[2x] + [2y] \geqslant [x + y].$$

Note that now $0 \leqslant [x + y] \leqslant 1$. If $[x + y] = 0$, then the result clearly holds. If $[x + y] = 1$, then $x + y \geqslant 1$, and at least one of x and y is greater or equal to $\frac{1}{2}$, says $x \geqslant \frac{1}{2}$. Hence $[2x] + [2y] \geqslant [2x] = 1$, it means (3.7) holds. We complete the proof.

Exercises

3.1 Prove that for any given positive integer $n > 1$, there exist n consecutive composite numbers.

3.2 Prove that there are infinitely many primes with the form $4k - 1$, also with the form $6k - 1$ (k is a positive integer).

3.3 Prove that there are infinitely many odds m, such that $8^m + 9m^2$ is composite.

3.4 Assume that integers a, b, c and d satisfy

$$a > b > c > d > 0$$

and

$$a^2 + ac - c^2 = b^2 + bd - d^2.$$

Prove that $ab + cd$ is not prime.

3.5 Prove the result in Example 6 of Chapter 2 by using the method in Example 4 of this Chapter.

4   Indeterminate Equations (I)

Indeterminate equations are equations in which the number of unknowns are greater than that of equations, and the range of unknowns are restricted (e.g., integers, positive integers, positive rational numbers, etc.). Indeterminate equation(s) is an important topic in number theory. They also occur in mathematical competitions.

In elementary mathematics, three main methods we often use in dealing with indeterminate equations are factorization, congruence, and estimation (of inequalities), among them factorization is the most important method.

Roughly speaking, the main function of factorization method is through "factorization" we can factor the original equations into some other equations which can be solved easily. Here, "factorization" includes two aspects of the way: firstly, it is a factorization for algebraic expressions. Secondly, it supplies some suitable decompositions by using some properties of integers (unique factorization theorem, relatively prime properties).

Of course, there is no fixed procedure to follow for factorization method. Sometimes factorization is very difficult or there are too many ways of factorization to choose. Sometimes, further proof is difficult. Examples in this chapter shows all these cases.

We often combine the factorization method with other ways, please refer to examples of this chapter and following chapters.

Example 1   If a positive integer, after adding 100, becomes a perfect square, and after adding 168, becomes another perfect square. Find this number.

Solution   Let x be this number. According to the assumptions, there are positive integers y and z, such

$$\begin{cases} x + 100 = y^2, \\ x + 168 = z^2. \end{cases}$$

Eliminate x from the above two equations, we get

$$z^2 - y^2 = 68.$$

Factorize the left side of the equation above, write the right side in the standard factorization form, and we have

$$(z - y)(z + y) = 2^2 \times 17. \tag{4.1}$$

Since $z - y$ and $z + y$ are all positive integers, and $z - y < z + y$, thus by (4.1) and the unique factorization theorem (see property (5) in Chapter 3) we have

$$\begin{cases} z - y = 1, \\ z + y = 2^2 \times 17; \end{cases} \quad \begin{cases} z - y = 2, \\ z + y = 2 \times 17; \end{cases} \quad \begin{cases} z - y = 2^2, \\ z + y = 17. \end{cases} \tag{4.2}$$

Solve these systems of linear equations with two unknowns, we get $y = 16$, $z = 18$. Therefore, $x = 156$.

Example 2   Find all integer solutions of the following indeterminate equation:

$$x^4 + y^4 + z^4 = 2x^2 y^2 + 2y^2 z^2 + 2z^2 x^2 + 24.$$

Solution   The key step (also the main difficulty of this problem) is to find out whether the equation can be factored into

$$(x + y + z)(x + y - z)(y + z - x)(z + x - y)$$
$$= -2^3 \times 3. \tag{4.3}$$

Since four divisors of the left side of the above equation are all integers, similar to Example 1, and by the unique factorization theorem, we can factor (4.3) into some (four unknowns linear) equations to solve. Although it works, it is more troublesome.

We use the following approach (based on (4.3)): since prime 2 divides the right side of (4.3), among the four divisors in the left side of (4.3) there is at least one divisor which is divisible by 2. On the other hand, sums of any two of these four numbers are all even, so they have the same parity. Thus they are all even, which implies that the left side of (4.3) is divisible by 2^4, but the right side of (4.3) is not a multiple of 2^4. Hence there is no solution for the original equation.

Incidentally, if in the solution of Example 1 we use the same way as Example 2, we can simplify the procedure slightly: since $z - y$ and $z + y$ have same parity, we just need to solve the second equation listed in (4.2).

About the second half of the proof of Example 2, if we consider it from the congruence (discussed in Chapter 6) point of view, it is clearer: firstly, consider (4.3) by modular 2, and then by modular 2^4. We will discuss specially the congruence method to solve indeterminate equations in Chapter 9.

Example 3 Prove that the product of two consecutive positive integers is neither a perfect square nor a perfect cube.

Proof Argue by contradiction. Suppose there are positive integers x and y, such that

$$x(x + 1) = y^2.$$

Multiple two sides of the equation by 4, and by rearrangement we get $(2x + 1)^2 = 4y^2 + 1$, which can be factored into

$$(2x + 1 + 2y)(2x + 1 - 2y) = 1.$$

Since two divisors of the left side are all positive integers, we have

$$\begin{cases} 2x + 1 + 2y = 1, \\ 2x + 1 - 2y = 1. \end{cases}$$

This gives solution $x = y = 0$, a contradiction. We complete the proof of the first statement.

However, for the equation

$$x(x + 1) = y^3,$$

the above factorization method does not work easily. We apply another factorization (based on the properties of numbers): If the given equation has positive integer solutions x and y, since x and $x + 1$ are relatively prime, and their product is a perfect cube, then x and $x + 1$ are all perfect cubes. Thus

$$x = u^3, \ x + 1 = v^3, \ y = uv,$$

where u and v are positive integers. It means $v^3 - u^3 = 1$, so

$$(v - u)(v^2 + uv + u^2) = 1,$$

but this is impossible. It is easy to see that by a similar argument we can prove that the product of any two consecutive positive integers is not k-th power of some integer (for some $k \geqslant 2$).

Discriminating whether divisors in a product are relatively prime is very important. Please see the following Example 4 and Example 5.

Example 4 Prove that the equation

$$y + y^2 = x + x^2 + x^3$$

has no integer solutions with $x \neq 0$.

Proof Assume that the equation has integer solution with $x \neq 0$, factor the equation into

$$(y - x)(y + x + 1) = x^3. \tag{4.4}$$

At first, we prove $\gcd(y - x, \ y + x + 1) = 1$. If it is not true, then there is a prime p which is a common divisor of $y - x$ and $y + x + 1$. By (4.4), $p \mid x^3$, so p divides x. Combining with $p \mid (y - x)$, we know $p \mid y$. But $p \mid (x + y + 1)$, thus $p \mid 1$, which is impossible. Therefore, the two divisors in the left side of (4.4) are relatively prime. But the right side of (4.4) is a perfect cube, thus there are integers a and b, such that

$$y - x = a^3, \ y + x + 1 = b^3, \ x = ab.$$

Eliminate x and y, and we have

$$b^3 - a^3 = 2ab + 1. \tag{4.5}$$

Now we prove that equation (4.5) has no integer solutions, which induces a contradiction. We factor (4.5) into

$$(b-a)(b^2+ab+a^2) = 2ab + 1. \qquad (4.6)$$

Note that $x = ab$ and $x \neq 0$, thus $ab \neq 0$. If $ab > 0$, then from (4.6) we get $b - a > 0$. Since a and b are integers, $b - a \geqslant 1$. Thus, the left side of (4.6) $\geqslant b^2 + ab + a^2 > 3ab >$ the right side. If $ab < 0$, then $|b-a| \geqslant 2$, so the absolute value of the left side of (4.6) $\geqslant 2(a^2 + b^2 - |ab|) > 2|ab|$, but the absolute value of the right side $< 2|ab|$. Therefore, (4.6) is false. It means there is no integer solution with $x \neq 0$ for the equation in the problem.

In the proof of equation (4.6) we have used inequality to estimate (the absolute value of the left side is always great than that of the right side). This method is known as estimation method. The estimation method (in number theory) is always based on integers. We can use various properties of integers and produce some suitable inequalities. For instance, in the above proof we apply one of the most elementary property of integers, that is, if integer $x > 0$, then $x \geqslant 1$.

Of course, using estimation method is not restricted to indeterminate equations. In many problems concerning number theory we can use this method. There are a lot of such examples in this book.

Example 5　Let k be a given positive integer, $k \geqslant 2$. Prove that

(1) the product of three consecutive integers is not a k-th power of some integers;

(2) the product of four consecutive integers is not a k-th power of some integers, too.

Proof　(1) Assume that there are integers $x \geqslant 2$ and y, such that

$$(x-1)x(x+1) = y^k. \qquad (4.7)$$

Note that the three divisors $x-1$, x and $x+1$ in the left side of the above equation are not always relatively prime pairwise, thus we cannot get the result that they are all k-th power of some integers. One method to overcome this difficulty is to rearrange equation (4.7) into

$$(x^2-1)x = y^k. \qquad (4.8)$$

Since x and $x^2 - 1$ are relatively prime, from (4.8) we obtain that there are positive integers a and b, such that

$$x = a^k, \ x^2 - 1 = b^k, \ ab = y.$$

Thus,

$$1 = a^{2k} - b^k = (a^2)^k - b^k$$
$$= (a^2 - b)(a^{2k-2} + a^{2k-4}b + \cdots + a^2 b^{k-2} + b^{k-1}).$$

Since $x \geqslant 2$, so $a \geqslant 2$. Also, $k \geqslant 2$, thus the second divisor in the above equation is great than 1, a contradiction.

(2) Suppose that there are positive integers x and y, such that

$$(x - 1)x(x + 1)(x + 2) = y^k. \tag{4.9}$$

The four divisors in the left side of the above equation are not always relatively prime pairwise, we classify it into two cases to determine whether they are relatively prime.

(i) x is odd. In this case, $\gcd(x, x + 2) = 1$. Since two consecutive integers are relatively prime, thus x and $x - 1$ are relatively prime, so are x and $x + 1$. Therefore, x and $(x - 1)(x + 1)(x + 2)$ are relatively prime. Due to (4.9), there are positive integers a and b, such that

$$x = a^k, \ (x - 1)(x + 1)(x + 2) = b^k.$$

We will prove that if $x \geqslant 2$ and $k \geqslant 3$, then

$$(x - 1)(x + 1)(x + 2) = x^3 + 2x^2 - x - 2$$

lies between the k-th powers of two consecutive integers, thus it cannot be a k-th power of some integer. In fact,

$$(a^3)^k = x^3 < x^3 + 2x^2 - x - 2 < x^3 + kx^2 + 1$$
$$= a^{3k} + ka^{2k} + 1 < a^{3k} + ka^{3(k-1)} + 1$$
$$< (a^3 + 1)^k \ (\text{by the binomial theorem}).$$

(ii) x is even. Similar to (i), we know $x + 1$ and $(x - 1)x(x + 2)$ are relatively prime. Thus, there are positive integers a and b, such that

$$x + 1 = a^k, \ (x-1)x(x+2) = b^k.$$

We will prove that if $k \geqslant 3$, then $(x-1)x(x+2)$ lies between the k-th powers of two consecutive integers.

On the one hand, it is easy to know

$$(x-1)x(x+2) < (x-1)(x+1)^2 < (x+1)^3 = (a^3)^k.$$

On the other hand, when $k = 3$ we have

$$(x-1)x(x+2) > x^3,$$

except for $x = 2$; but when $x = 2$ the statement is clearly true. Thus when $k = 3$ the result holds.

For $k \geqslant 4$, we have

$$\begin{aligned}
(x-1)x(x+2) &= (a^k - 2)(a^k - 1)(a^k + 1) \\
&= a^{3k} - 2a^{2k} - a^k + 2 \\
&> a^{3k} - ka^{2k} \\
&= ((a^3 - 1) + 1)^k - ka^{2k} \\
&> (a^3 - 1)^k + k(a^3 - 1)^{k-1} - ka^{2k} \\
&> (a^3 - 1)^k.
\end{aligned}$$

To see the last step is true, we note that since x is even, thus a is odd and $a \geqslant 3$.

When $a = 3$ it is clearly true. When $a \geqslant 5$, since $k \geqslant 4$, we have

$$(a^3 - 1)^{k-1} = (a-1)^{k-1}(a^2 + a + 1)^{k-1} > (a-1)^3 a^{2(k-1)} > a^{2k}.$$

Combine with (i) and (ii) and we prove that if $k \geqslant 3$ the product of four consecutive positive integers is not a k-th power of some integer. When $k = 2$ the proof for the result is very easy, and we leave it to the readers.

Exercises

4.1 Prove that the product of four consecutive positive integers is not a perfect square.

4.2 Find all integers which can be expressed as the difference of two squares of some integers.

4.3   Find all integer solutions for the indeterminate equation

$$\begin{cases} x + y + z = 3, \\ x^3 + y^3 + z^3 = 3. \end{cases}$$

Selected Lectures on Competition Problems (I)

From the previous chapters we can find out a notable feature in number theory, that is, flexible and diverse, particularly for problems in number theory in Mathematics Olympiad. In this chapter we choose more such examples.

Example 1   Suppose $m \geqslant n \geqslant 1$, prove that $\dfrac{\gcd(m, n)}{m} \dbinom{m}{n}$ is an integer.

Proof   When $x = m$, the value of $\dfrac{x}{m} \dbinom{m}{n}$ is $\dbinom{m}{n}$ which is an integer. When $x = n$, it is

$$\frac{n}{m} \cdot \frac{m}{n} \binom{m-1}{n-1} = \binom{m-1}{n-1}$$

which is also an integer. By Bézout's identity, there are integers u and v, such that

$$\gcd(m, n) = mu + nv.$$

Hence,

$$\frac{\gcd(m, n)}{m} \binom{m}{n} = u \binom{m}{n} + v \frac{n}{m} \binom{m}{n}$$

is an integer.

Example 2   Assume that a and b are two different positive integers, and $ab(a + b)$ is a multiple of $a^2 + ab + b^2$. Prove that $|a - b| > \sqrt[3]{ab}$.

Proof   From the given conditions and the identity

$$a(a^2 + ab + b^2) - ab(a + b) = a^3,$$

we know $(a^2 + ab + b^2) \mid a^3$. Similarly, $(a^2 + ab + b^2) \mid b^3$, so $(a^2 + ab + b^2) \mid \gcd(a^3, b^3)$.

Thus

$$(a^2 + ab + b^2) \mid \gcd(a, b)^3. \tag{5.1}$$

Set $d = \gcd(a, b)$, $a = a_1 d$, $b = b_1 d$, then (5.1) is equivalent to $(a_1^2 + a_1 b_1 + b_1^2) \mid d$. Thus, $d \geqslant a_1^2 + a_1 b_1 + b_1^2$, and $d > a_1 b_1$. Since $a \neq b$, integers $a_1 \neq b_1$. Hence $\mid a_1 - b_1 \mid \geqslant 1$, and we get

$$\mid a - b \mid^3 = d^3 \mid a_1 - b_1 \mid^3 \geqslant d^3 > d^2 a_1 b_1 = ab,$$

that is, $\mid a - b \mid > \sqrt[3]{ab}$.

In the proof of Example 2, firstly, from the properties of integers such as divisibility we get divisibility relation (5.1), then use the inequality (property (3) in Chapter 1). This is a basic way to deal with inequality problems of integers and to solve problems of number theory using the estimation method. The following two examples are all done in this way.

Example 3   Choose arbitrarily several different integers between two adjacent perfect squares n^2 and $(n + 1)^2$, prove that the products of any two of the integers are not equal pairwise.

Proof   Assume that integers a, b, c and d satisfy

$$n^2 < a < b < c < d < (n + 1)^2.$$

Clearly, we just need to prove that $ad \neq bc$. We shall prove by contradiction. Assume that the above numbers a, b, c and d satisfy $ad = bc$. Then by proof one of Example 3 in Chapter 3 we know, there are positive integers p, q, u and v, such that

$$a = pu, \ b = qu, \ c = pv, \ d = qv.$$

By $b > a$ and $c > a$ we get $q > p$ and $v > u$. Since p, q, u and v are integers, $q \geqslant p + 1$, and $v \geqslant u + 1$. Thus (note that $a = pu > n^2$)

$$d = qv \geqslant (p + 1)(u + 1) = pu + (p + u) + 1$$

$$> n^2 + 2\sqrt{pu} + 1 > n^2 + 2n + 1 = (n+1)^2,$$

a contradiction.

Example 4   Find all positive integer solutions of the indeterminate equation

$$(n-1)! = n^k - 1. \tag{5.2}$$

Solution   When $n = 2$, from (5.2) we get $(n, k) = (2, 1)$. If $n > 2$, the left side of (5.2) is even, so the right side is also even, thus n is odd. When $n = 3$ or 5, we solve (5.2) and get solutions $(n, k) = (3, 1)$, $(5, 2)$.

In the following discussion we assume $n > 5$ and n is odd. In this case $\dfrac{n-1}{2}$ is an integer and $\dfrac{n-1}{2} < n - 3$, so $2 \cdot \dfrac{n-1}{2} \Big| (n-2)!$, that is, $(n-1) \mid (n-2)!$. Hence, $(n-1)^2 \mid (n-1)!$, that is to say,

$$(n-1)^2 \mid (n^k - 1). \tag{5.3}$$

On the other hand, by the binomial theorem, we have

$$n^k - 1 = ((n-1) + 1)^k - 1$$
$$= (n-1)^k + \binom{k}{1}(n-1)^{k-1} + \cdots + \binom{k}{k-2}(n-1)^2 + k(n-1). \tag{5.4}$$

From (5.3) and (5.4) we get $(n-1)^2 \mid k(n-1)$, that is, $(n-1) \mid k$. So $k \geqslant n - 1$.

Thus

$$n^k - 1 \geqslant n^{n-1} - 1 > (n-1)!.$$

It means that when $n > 5$ equation (5.2) has no positive integer solutions. Thus, all positive integer solutions of (5.2) are $(n, k) = (2, 1), (3, 1), (5, 2).$

Example 5   Find all integer solutions of the equation

$$x^3 + x^2 y + x y^2 + y^3 = 8(x^2 + xy + y^2 + 1).$$

Solution One   The left side of the original equation is a cubic polynomial function of x and y. For integers x and y, the absolute values of the cubic polynomial are usually greater than those of quadratic polynomials. So maybe we can use the estimation method to solve the equation. We factorize the equation into

$$(x^2 + y^2)(x + y - 8) = 8(xy + 1). \tag{5.5}$$

If $x + y - 8 \geqslant 6$, then $x + y \geqslant 14$. Thus

$$x^2 + y^2 \geqslant \frac{(x + y)^2}{2} > 4.$$

Now the left side of (5.5)

$$\geqslant 6(x^2 + y^2) = 4(x^2 + y^2) + 2(x^2 + y^2)$$
$$\geqslant 8xy + 2(x^2 + y^2) > 8(xy + 1),$$

so in this case the equation has no integer solutions.

If $x + y - 8 \leqslant -4$, then $x + y \leqslant 4$, in this case the left side of (5.5)

$$\leqslant -4(x^2 + y^2) \leqslant -4 \times 2 \mid xy \mid \leqslant 8xy < 8(xy + 1).$$

This time the equation has no integer solutions, too. Therefore, the integer solutions of the equation (x, y) satisfy

$$-3 \leqslant x + y - 8 \leqslant 5.$$

On the other hand, the left side of (5.5) is even, which implies that x and y have the same parity. Thus $x + y - 8$ is even, it is -2, 0, 2 or 4. Combining with (5.5) and by checking we can easily get all solutions $(x, y) = (2, 8), (8, 2)$.

Solution Two   Set $u = x + y$, $v = xy$. Then the original equation can be rewritten as

$$u(u^2 - 2v) = 8(u^2 - v + 1), \tag{5.6}$$

or equivalently,

$$u^3 - 2uv = 8u^2 - 8v + 8,$$

which implies that u is even. Let $u = 2w$. Then

$$2w^3 - vw = 8w^2 - 2v + 2. \tag{5.7}$$

We solve it and get

$$v = \frac{2w^3 - 8w^2 - 2}{w - 2} = 2w^2 - 4w - 8 - \frac{18}{w-2}. \tag{5.8}$$

Hence, $w - 2$ is a divisor of 18, that is, one of ± 1, ± 2, ± 3, ± 6, ± 9, ± 18. For every possible value of w, combining with (5.8) we can determine the value of v, and get the corresponding integer solutions (x, y) which are only $(2, 8)$ and $(8, 2)$. (Note that after finding a pair of values of w and v the condition that x and y are integers is equivalent to that $w^2 - v$ is a perfect square.)

Remark   The two sides of the original equation are all symmetric polynomials with two variables x and y which can be expressed as polynomials of $u = x + y$ and $v = xy$ (cf. (5.6)). The feature of this expression is that the derived equation (5.7) is a linear equation of v, and we can solve for v (as an expression of w).

In Example 6 below, we make use of certain expression of numbers.

Example 6   Find positive integer(s) n such that it is divisible by all positive integers which are less than or equal to $\sqrt{n}$.

Solution   Firstly, we prove that every positive integer n can be uniquely expressed in the form

$$n = q^2 + r, \ 0 \leqslant r \leqslant 2q. \tag{5.9}$$

This is because any positive integer n must be between two adjective perfect squares, that is, there is positive integer q, such that

$$q^2 \leqslant n < (q+1)^2.$$

Set $r = n - q^2$, then $r \geqslant 0$, and

$$r < (q+1)^2 - q^2 = 2q + 1.$$

So the integer $r \leqslant 2q$, thus n has an expression of the form (5.9).

On the other hand, if n can be expressed in the form (5.9), then $q^2 \leqslant n < (q+1)^2$. So $q = [\sqrt{n}]$, it means that q is uniquely determined by n. Thus r is also determined.

It is not difficult to solve Example 6 by making use of (5.9).

From the given condition $q = [\sqrt{n}]$ divides n, combining with (5.9) we know $q \mid r$, so $r = 0$, q or $2q$, that is, n has the forms

$$n = q^2, \ q^2 + q, \ q^2 + 2q.$$

Now when $n = 1$, 2, 3, the given condition is clearly satisfied. Let $n > 3$, then $q = [\sqrt{n}] \geqslant 2$. So from the given condition we have $(q - 1) \mid n$. If $n = q^2$, from

$$q^2 = q(q - 1) + q \text{ and } \gcd(q - 1, q) = 1,$$

we have $q - 1 = 1$, or $q = 2$. Hence $n = 4$.

Similarly, if $n = q^2 + q$, then $q = 2$, 3. Thus $n = 6$, 12. If $n = q^2 + 2q$, then $q = 2$ or 4, and $n = 8$, 24. Therefore, all possible values of n are 1, 2, 3, 4, 6, 8, 12 and 24. By checking, they all satisfy the required condition.

Example 7   Prove that among any 51 numbers chosen randomly from 1, 2, ..., 100 there are two numbers which are relatively prime.

Proof   When someone points out the key step in solving the problem you will find it very simple. From 1, 2, ..., 100 we choose two consecutive numbers in the sequential order, and put them into 50 pairs

$$\{1, 2\}, \{3, 4\}, \ldots, \{99, 100\}.$$

Then any 51 numbers chosen randomly must contain one of the pairs above. Since the two numbers are consecutive, of course they are relatively prime.

Example 8   Prove that there exist 1000 consecutive positive integers, such that there are exactly 10 primes among them.

Proof   The basis of this proof is Exercise 3.1. From the result in the exercise we know there are 1000 consecutive positive integers

$$a, a + 1, \ldots, a + 999, \tag{5.10}$$

among them no number is prime.

Now we do the following operation to the numbers in (5.10): deleting the far right number $a + 999$ and adding $a - 1$ on the far left.

Clearly, in the resulting sequence

$$a - 1, a, \ldots, a + 998$$

there is at most one prime. Repeat this operation, until we reach 1, 2, $\ldots$, 1000. We note that the number of primes among (consecutive 1000) positive numbers obtained after an operation is the same, increases by one, or decreases by one, comparing with the number of primes among these numbers before the operation. Clearly, there are more than 10 primes among the finally obtained numbers 1, 2, $\ldots$, 1000. Hence among these operations above there must exist one operation after which there are exactly 10 primes among the 1000 consecutive numbers obtained.

Example 7 and Example 8 are all so-called "existence problems", it means that to prove that "something" possesses "some property", and in the proof we actually do not construct the required property, but using a logical arsument we can show that it exists. In Example 7 we apply the well-known "pigeonhole principle", and in Example 8 we apply the following principle, which is sometimes called "discrete nullsatz":

Let $f(x)$ be a function defined on a (positive) integer set, and its values are also integers. If $| f(n) - f(n + 1) | \leqslant 1$ for all n, and there are integers a and b, such that $f(a)f(b) < 0$, then there exists an integer c between a and b, such that $f(c) = 0$. (In Example 8 we can take $g(n)$ the number of primes among the consecutive 1000 positive integers starting with n, and $f(n) = g(n) - 10$.)

Another effective method other than existence proof is constructive method, i. e., actually constructing the required property. Constructive method is an important method, it has various forms. In number theory many problems have to be proved by constructive method. In what follows, we give some examples.

Example 9　For a positive integer, if every power of a prime divisor in its unique prime factorization is great than 1, we call the integer a power number. Prove that there exist infinitely many different positive integers, such that they and sums of any finite number of different integers in them are not power numbers.

Proof   Let $2 = p_1 < p_2 < \cdots < p_n < \cdots$ be all primes. Then

$$p_1, \ p_1^2 p_2, \ p_1^2 p_2^2 p_3, \ \dots, \ p_1^2 p_2^2 \cdots p_{n-1}^2 p_n, \ \dots \qquad (5.11)$$

are required numbers.

To verify this statement, we denote by a_n the n-th number in the above sequence. At first, every a_n is not a power number. For any r, $s, \dots, n \ (1 \leqslant r < s < \cdots < n)$, by (5.11) we have $p_r \mid a_r$, but $p_r^2 \nmid a_r$, and $p_r \left| \dfrac{a_s}{a_r} \right., \dots, p_r \left| \dfrac{a_n}{a_r} \right.$. Hence, in

$$a_r + a_s + \cdots + a_n = a_r \left(\frac{a_s}{a_r} + \cdots + \frac{a_n}{a_r} + 1 \right),$$

the second divisor is relatively prime with p_r. Thus the prime p_r just occurs once in the unique prime factorization of $a_r + a_s + \cdots + a_n$, and $a_r + a_s + \cdots + a_n$ is not a power number. On the other hand, since primes are infinitely many, the numbers in (5.11) are also infinitely many.

Example 10   Prove that there are infinitely many positive numbers n such that $n \mid (2^n + 1)$.

Proof One   Let us consider the first values of n, among numbers less than 10 there are only $n = 3^0, \ 3^1, \ 3^2$ which satisfy the requirement. We expect all $n = 3^k \, (k \geqslant 0)$ satisfy the requirement.

Proving it is a simple exercise by induction. The first step is clearly true. Suppose that for $k \geqslant 0$ we have $3^k \mid (2^{3^k} + 1)$, it means

$$2^{3^k} = -1 + 3^k u, \text{ for some integer } u.$$

Then

$$2^{3^{k+1}} = (-1 + 3^k u)^3 = -1 + 3^{k+1} v \text{ (for some integer } v),$$

so $3^{k+1} \mid (2^{3^{k+1}} + 1)$, which implies that $n = 3^{k+1}$ satisfy the requirement too. By induction, we finish the proof.

Proof Two   This is a different constructive method. The key is to note that if $n \mid (2^n + 1)$, then $m \mid (2^m + 1)$ for $m = 2^n + 1$.

In fact, as $2^n + 1$ is odd, if $2^n + 1 = nk$ for some integer k, then k is odd. Thus

$$2^m + 1 = (2^n)^k + 1 = (2^n + 1)((2^n)^{k-1} - (2^n)^{k-2} + \cdots - 2^n + 1)$$

is a multiple of $m = 2^n + 1$.

By this result, we can recursively get infinitely many required numbers: $1, 3, 9, 513, \ldots$.

Solutions by the above two methods are not completely the same, but they are all multiples of 3, except for the number 1. This is not accidental. In fact, by Example 1 in Chapter 8 every required integer $n \; (> 1)$ is divisible by 3.

Example 11 Prove that there are infinitely many positive integers such that $n \mid (2^n + 2)$.

Proof This problem seems to be so similar to Example 10, but actually it is more difficult. We still use constructive method inductively, and the key is to strengthen the inductive assumption. In the following we show that if n satisfies

$$2 \mid n, \; n \mid (2^n + 2), \; (n - 1) \mid (2^n + 1), \qquad (5.12)$$

then for $m = 2^n + 2$, we have

$$2 \mid m, \; m \mid (2^n + 2), \; (m - 1) \mid (2^n + 1). \qquad (5.13)$$

Since $2^n + 2 = 2(2^{n-1} + 1)$ is a product of 2 and an odd number and $2 \mid n$, the integer k in $2^n + 2 = nk$ is odd, thus

$$2^m + 1 = 2^{nk} + 1 = (2^n)^k + 1$$

is a multiple of $2^n + 1 = m - 1$.

Similarly, from $2^n + 1 = (n - 1)l$ we know l is odd, so

$$2^m + 2 = 2(2^{m-1} + 1) = 2((2^{n-1})^l + 1)$$

is a multiple of

$$2(2^{n-1} + 1) = 2^n + 2 = m.$$

Clearly, $m = 2^n + 2$ is even. Thus the above assertion be proved.

Now since $n = 2$ satisfies (5. 12), by (5. 13) we can construct recursively infinitely many required numbers, such as 2, 6, 66, ...

We note that in (5. 12) $2 \mid n$ is necessary, that is, the numbers satisfying the requirement of this problem are all even. Since if there is an odd $n > 1$ such that $n \mid (2^n + 2)$, then $n \mid (2^{n-1} + 1)$, which contradicts the conclusion in Example 3 of Chapter 8.

Exercises

5. 1   Let $\gcd(m, n) = 1$. Prove that $m \mid \binom{m + n - 1}{n}$.

5. 2   Prove that a positive integer n can be expressed as the sum of some (at least two) consecutive positive integers if and only if n is not a power of 2.

5. 3   Prove that every positive integer n can be expressed in the form of $a - b$, where a and b are positive integers, such that the numbers of distinct prime divisors of a and b are the same.

5. 4   Prove that the equation $x! \cdot y! = z!$ has infinitely many positive integer solutions x, y and z with $x < y < z$.

5. 5   Prove that for every $n \geqslant 2$ there are n distinct positive integers $a_1, a_2, \ldots, a_n$, such that $(a_i - a_j) \mid (a_i + a_j)(1 \leqslant i, j \leqslant n, i \neq j)$.

6   Congruence

Congruence is an important concept in number theory. It has a wide range of applications.

Let n be a positive integer. If integers a and b satisfy $n \mid (a - b)$, then they are said to be congruent modulo n, and written

$$a \equiv b \pmod{n}.$$

If $n \nmid (a - b)$, then a and b are said to be not congruent modulo n, and written

$$a \not\equiv b \pmod{n}.$$

By the division algorithm, a and b are congruent modulo n if and only if a and b give the same remainder when they are divided by n.

For a fixed modulo n, congruence modulo n has the same characteristics as equality.

(1) (Reflexivity) $a \equiv a \pmod{n}$.

(2) (Symmetry) If $a \equiv b \pmod{n}$, then $b \equiv a \pmod{n}$.

(3) (Transitivity) If $a \equiv b \pmod{n}$, and $b \equiv c \pmod{n}$, then $a \equiv c \pmod{n}$.

(4) (Addition of congruences) If $a \equiv b \pmod{n}$, and $c \equiv d \pmod{n}$, then $a \pm c \equiv b \pm d \pmod{n}$.

(5) (Multiplication of congruences) If $a \equiv b \pmod{n}$, and $c \equiv d \pmod{n}$, then $ac \equiv bd \pmod{n}$.

It is easy to see that by applying (4) and (5) repeatedly, we can get operation formulas of addition and multiplication for more than two congruences (with same modulo). In particular, from (5) we can easily get, if $a \equiv b \pmod{n}$, k and c are integers with $k > 0$, then

$$a^k c \equiv b^k c \pmod{n}.$$

Note that, in general, the elimination law is not true for congruence, that is, from $ac \equiv bc \pmod{n}$ we cannot always get $a \equiv b \pmod{n}$. However, we have the following result:

(6) If $ac \equiv bc \pmod{n}$, then $a \equiv b \left(\bmod \dfrac{n}{\gcd(n, c)}\right)$, which implies that if $\gcd(c, n) = 1$, then $a \equiv b \pmod{n}$, that is, when c and n are relatively prime, we can eliminate c in the two sides of congruence and do not change its modulo (this once again demonstrates the importance of the relatively prime property).

Now we list some simple but useful properties concerning modulo.

(7) If $a \equiv b \pmod{n}$ and $d \mid n$, then $a \equiv b \pmod{d}$.

(8) If $a \equiv b \pmod{n}$ and $d \neq 0$, then $da \equiv db \pmod{dn}$.

(9) If $a \equiv b \pmod{n_i}$, $i = 1, 2, \ldots, k$, then $a \equiv b \pmod{[n_1, n_2, \ldots, n_k]}$. In particular, if $n_1, n_2, \ldots, n_k$ are relatively prime pairwise, then $a \equiv b \pmod{n_1 n_2 \cdots n_k}$.

From the above properties (1), (2) and (3), an integer set can be classified via modulo n. More precisely, if a and b are congruent modulo n, then they belongs to the same class, otherwise they belong to different classes. Every such class is called a congruent class modulo n.

By the division algorithm, any integer is congruent exactly to one of the numbers $0, 1, \ldots, n - 1$ modulo n, and the n numbers $0, 1, \ldots, n - 1$ are not congruent to each other modulo n. Therefore there are totally n different classes modulo n, they are

$$M_i = \{x \mid x \in \mathbf{Z}, x \equiv i \pmod{n}\}, \ i = 0, 1, \ldots, n - 1.$$

For instance, there are two congruent classes modulo 2, which are the class of even numbers and the class of odd numbers. Numbers in these two classes have the forms $2k$ and $2k + 1$, respectively (where k is any integer).

In every class among the n congruence classes we choose arbitrarily one number as representative. The set of these n numbers is called a complete system of residues modulo n, briefly called a complete system modulo n. In other words, set of n numbers c_1, c_2, ..., c_n is called a complete system modulo n if and only if any two of them are not congruent to each other modulo n. For example, 0, 1, ..., $n - 1$ is a complete system of residues modulo n. It is called the minimal complete system of residues modulo n.

It is easy to see that if i and n are relatively prime, then any number in the congruence class M_i is relatively prime with n. This kind of congruence class is called a reduced congruence class modulo n. We denote the number of the reduced congruence classes modulo n by $\varphi(n)$, and is called the Euler function. It is an important function in number theory. Clearly, $\varphi(1) = 1$, and for $n > 1$, $\varphi(n)$ is the number of elements among 1, 2, ..., $n - 1$ which is relatively prime with n. For example, if p is a prime, then $\varphi(p) = p - 1$.

Choose arbitrarily a number in every $\varphi(n)$ reduced congruence class modulo n as a representative, these $\varphi(n)$ numbers form a reduced residue system modulo n, or in short a reduced system modulo n. Thus $\varphi(n)$ numbers r_1, r_2, ..., $r_{\varphi(n)}$ are called a reduced system modulo n if they are not congruent mutually and all relatively prime with n. $\varphi(n)$ positive integers which are less than n and relatively prime with n are called the minimal positive reduced system modulo n.

The following result, producing a new complete (or reduced) system modulo n from a complete (or reduced) system modulo n, has many applications.

(10) Let $\gcd(a, n) = 1$, b be any integer.

If c_1, c_2, ..., c_n is a complete system modulo n, $ac_1 + b$, $ac_2 + b$, ..., $ac_n + b$ is also a complete system modulo n.

If r_1, r_2, ..., $r_{\varphi(n)}$ is a reduced system modulo n, then ar_1, ar_2, ..., $ar_{\varphi(n)}$ is also a reduced system modulo n.

From the first statement in (10) we deduce

(11) Let $\gcd(a, n) = 1$ and b be any integer. Then there are integers x, such that $ax \equiv b \pmod{n}$, and all such x's form a congruence class modulo n.

In particular, there are x such that $ax \equiv 1 \pmod{n}$. Such x is called an inverse of a modulo n, and is denoted by a^* or $a^{-1} (\bmod n)$. They form a congruence class modulo n, therefore there is a^{-1} satisfying $1 \leqslant a^{-1} < n$.

We know that there are n possible values for the remainders of an integer modulo n. But for squares, cubes of integers, their numbers of remainders modulo n may be significantly reduced. This fact is a basic point in solving many problems by congruence method. The following simple conclusion has a wide range of applications.

(12) Perfect squares are congruent to 0 or 1 modulo 4, congruent to 0, 1, or 4 modulo 8, congruent to 0 or 1 modulo 3, congruent to 0 or ± 1 modulo 5.

Perfect cubes are congruent to 0 or ± 1 modulo 9.

The 4th powers of integers are congruent to 0 or 1 modulo 16.

Now we have a number of examples to illustrate the role of congruence in problem solving.

Example 1   Let a, b, c and d be positive integers. Prove that $a^{4b+d} - a^{4c+d}$ is divisible by 240.

Proof   Since $240 = 2^4 \times 3 \times 5$, we want to prove that $a^{4b+d} - a^{4c+d}$ is divisible by 3, 5 and 16, thus the result follows (cf. the Remark in Example 5, Chapter 3).

Firstly, we prove $3 \mid (a^{4b+d} - a^{4c+d})$. By (12) we have $a^2 \equiv 0, 1 \pmod{3}$, and $a^{4b} \equiv a^{4c} \equiv 0, 1 \pmod{3}$. Thus

$$a^{4b+d} - a^{4c+d} = a^d (a^{4b} - a^{4c}) \equiv 0 \pmod{3}.$$

Similarly, from $a^2 \equiv 0, \pm 1 \pmod{5}$ we get $a^4 \equiv 0, 1 \pmod{5}$, so $a^{4b} \equiv a^{4c} \equiv 0, 1 \pmod{5}$. Thus $a^{4b+d} - a^{4c+d} \equiv \pmod{5}$.

Finally, from $a^4 \equiv 0, 1 \pmod{16}$ we know $a^{4b} \equiv a^{4c} \equiv 0, 1 \pmod{16}$. Thus $a^{4b+d} - a^{4c+d} \equiv 0 \pmod{16}$. This completes the proof.

Example 1 is a routine problem. However, the following Example 2 needs some little techniques.

Example 2   Let a, b and c be integers such that $a + b + c = 0$. Set $d = a^{1999} + b^{1999} + c^{1999}$. Prove that $|d|$ is not prime.

Proof   There are many methods to solve this problem. We use congruence to prove that d has a nontrivial fixed divisor.

First, for any integer u, the numbers u^{1999} and u have the same parity, i.e., $u^{1999} \equiv u \pmod{2}$. So $d \equiv a + b + c \equiv 0 \pmod{2}$. Thus $2 \mid d$.

On the other hand, it is easy to verify (for the cases $3 \mid u$ and $3 \nmid u$)

$$u^3 \equiv u \pmod{3}, \tag{6.1}$$

which implies that

$$u^{1999} \equiv u \cdot u^{1998} \equiv u \cdot u^{666} \equiv u \cdot u^{222} \equiv u^{75}$$
$$\equiv u^{25} \equiv u^9 \equiv u^3 \equiv u \pmod{3}.$$

Hence, $d \equiv a + b + c \equiv 0 \pmod{3}$. Therefore $6 \mid d$, and d is not prime.

Remark The congruence (6.1) in the solution is a special case of famous Fermat's Little Theorem. Please refer to the next chapter.

Example 3 Assume that integers x, y and z satisfy

$$(x - y)(y - z)(z - x) = x + y + z. \tag{6.2}$$

Prove that $x + y + z$ is divisible by 27.

Proof From (6.2) we will deduce that x, y and z are mutually congruent modulo 3, thus $27 \mid (x - y)(y - z)(z - x)$. That is, $27 \mid (x + y + z)$.

We prove by contradiction. First, we assume that there are just two of the numbers x, y and z are congruent modulo 3. We can assume $x \equiv y \pmod{3}$, but $x \not\equiv z \pmod{3}$. In this case $3 \mid (x - y)$ but $3 \nmid (x + y + z)$. Thus, the left side of (6.2) $\equiv 0 \pmod{3}$, but the right side $\not\equiv 0 \pmod{3}$, a contradiction. Thus this case does not occur.

Next, assume that each two of x, y and z are not congruent modulo 3. In this case, it is easy to verify $3 \mid (x + y + z)$, but $3 \nmid (x - y)(y - z)(z - x)$. Hence the remainders in the two sides of (6.2) are different modulo 3, a contradiction. Thus this case is also impossible.

Therefore, our assertion above is true. The proof is complete.

Remark The solution in Example 3 embodies a fundamental principle in dealing with problems of number theory using congruence: if an integer $A = 0$, then the remainder of A divided by any positive integer n ($n > 1$) is also 0. Hence if we can find some $n > 1$ such that A is not congruent to 0 modulo n, then A is not 0. Usually we use this principle to obtain a necessary condition via congruence, to deduce the result (as Example 3), or to prepare for a further proof. There are more such examples in what follows.

The following is an old problem.

Example 4   Let $n > 1$. Prove that $\underbrace{11\cdots1}_{n}$ is not a perfect square.

Proof   We prove by contradiction. Assume that there is some $n > 1$ and an integer x such that

$$\underbrace{11\cdots1}_{n} = x^2. \tag{6.3}$$

By (6.3), x is odd (In fact, (6.3) modulo 2, and note that $x^2 \equiv x$ (mod 2)). Furthermore, $x \equiv 1$ (mod 4), since $2 \nmid x$. But

$$\underbrace{11\cdots1}_{n} - 1 = \underbrace{11\cdots1}_{n-1}0$$

is divisible by 2, not by 4. That means the left side of (6.3) $\not\equiv 1$ (mod 4), a contradiction!

The key in dealing with such problems is the choice of modulo n. But concerning how to choose modulo there is no simple rule. It depends on the specific question. In Example 4, we first calculate (6.3) by modulo 2. Though we cannot solve the problem, we get some useful information, and thereafter calculate it by modulo 4, and deduce a contradiction.

Example 5   Using digits 1, 2, 3, 4, 5, 6 and 7 we can get 7-digit numbers and every digit is used only once in every such 7-digit number. Prove that there is none of these numbers which is a multiple of another one.

Proof   Suppose that there are such two 7-digit numbers a and b ($a \neq b$) such that

$$a = bc, \tag{6.4}$$

where c is an integer great than 1. Since the sum of digits in a or b is

$1+2+3+4+5+6+7 \equiv 1 \pmod 9$, $a \equiv b \equiv 1 \pmod 9$ (cf. Remark 2 in Example 3, Chapter 1). Now by modulo 9 for (6.4), we get $c \equiv 1$ (mod 9). But $c > 1$, so $c \geqslant 10$, and $a \geqslant 10b > 10^7$. A contradiction, since a is a 7-digit number.

Example 6 Assume that the sequence $\{x_n\}$ is 1, 3, 5, 11, ... and meets the following recursive relation

$$x_{n+1} = x_n + 2x_{n-1}, \; n \geqslant 2. \tag{6.5}$$

The sequence $\{y_n\}$ is 7, 17, 55, 161, ... and meets the following recursive relation

$$y_{n+1} = 2y_n + 3y_{n-1}, \; n \geqslant 2. \tag{6.6}$$

Prove that these two sequences have no common terms.

Proof Consider modulo 8. First, we prove that the sequence $\{x_n\}$ modulo 8 is a periodic sequence

$$1, 3, 5, 3, 5, \ldots. \tag{6.7}$$

Since $x_2 \equiv 3$, $x_3 \equiv 5 \pmod 8$. If we have

$$x_{n-1} \equiv 3, \; x_n \equiv 5 \pmod 8,$$

then by recursive equation (6.5), we have

$$x_{n+1} = x_n + 2x_{n-1} \equiv 5 + 2 \times 3 \equiv 3 \pmod 8,$$
$$x_{n+2} = x_{n+1} + 2x_n \equiv 3 + 2 \times 5 \equiv 5 \pmod 8.$$

This proves inductively our assertion.

Similarly, by (6.6) we can show that the sequence $\{y_n\}$ modulo 8 is a periodic sequence

$$7, 1, 7, 1, 7, 1, \ldots. \tag{6.8}$$

From (6.6) and (6.7) we find that the two sequences x_2, x_3, ... and y_1, y_2, ... modulo 8 have no terms with the same value. Furthermore, since $\{y_n\}$ is increasing, y_1, y_2, ... cannot be the same as $x_1 = 1$, which implies that $\{x_n\}$ and $\{y_n\}$ have no terms with the same value.

Remark After modulo 8 linearly recursive sequences (6.5) and

(6.6) become periodic sequences. This happens not by chance. In fact, for a given $m > 1$, if $\{x_n\}$ $(n \geqslant 1)$ is an integer sequence determined by the recursive equation

$$x_{n+k} = f(x_{n+k-1}, \ldots, x_{n+1}, x_n),$$

where f is a polynomial with integer coefficients and k unknowns, and the initial values $x_1, x_2, \ldots, x_k$ are given integers. Then after some terms, $\{x_n\}$ modulo m is a periodic sequence.

In order to prove this assertion, we denote by $\bar{x}_i$ the remainder of x_i divided by m $(0 \leqslant \bar{x}_i < m)$. Consider the ordered k-tulpes

$$A_n = \langle \bar{x}_n, \bar{x}_{n+1}, \ldots, \bar{x}_{n+k-1} \rangle (n = 1, 2, \ldots).$$

Since every $\bar{x}_i$ has at most m different values, there are at most m^k different A_n. Hence among $m^k + 1$ k-tuples $A_1, A_2, \ldots, A_{m^k+1}$ there are two of them which are exactly the same. Let us say $A_i = A_j (i < j)$, i.e.,

$$\bar{x}_{i+t} = \bar{x}_{j+t}(t = 0, 1, \ldots, k-1).$$

Combine with the recursive formula of $\{x_n\}$ and basic properties of congruence, and we deduce that the above equation still holds when $t = k$, i.e., $\bar{x}_{i+k} = \bar{x}_{j+k}$. Thus we can prove inductively, for any $t \geqslant 0$, $\bar{x}_{i+t} = \bar{x}_{j+t}$, which means that staring from the i-th term $\{x_n\}$ will occur periodically with every $j - i$ terms as a block.

Example 7   Let p be a given positive integer. Determine the minimum positive value of $(2p)^{2m} - (2p-1)^n$, where m and n are any positive integers.

Solution   The required minimum positive value is

$$(2p)^2 - (2p-1)^2 = 4p - 1.$$

For proving it we first note that, by

$$(2p)^2 = (4p-2)p + 2p$$

and

$$(2p-1)^2 = (4p-2)(p-1) + (2p-1),$$

we can easily get

$$(2p)^{2m} - (2p-1)^n \equiv (2p) - (2p-1) \equiv 1 \pmod{4p-2}.$$
$$(6.9)$$

Furthermore, we have to prove that there are no positive integers m and n such that $(2p)^{2m} - (2p-1)^n = 1$. If not, then

$$((2p)^m - 1)((2p)^m + 1) = (2p-1)^n.$$

The two divisors of the left side of the above equation are clearly relatively prime, but the right side is the n-th power of some positive integer. So

$$(2p)^m + 1 = a^n,$$
$$(6.10)$$

where a is a positive integer, and $a \mid (2p-1)$. Make equation (6.10) modulo a, then the left side is

$$(2p-1+1)^m + 1 \equiv 1 + 1 \equiv 2 \pmod{a},$$

which implies that $2 \equiv 0 \pmod{a}$. But clearly a is an odd integer greater than 1, a contradiction.

Combining with (6.9) we know if

$$(2p)^{2m} - (2p-1)^n > 0$$

then

$$(2p)^{2m} - (2p-1)^n \geqslant 4p-1,$$

and when $m = 1$ and $n = 2$ the equal sign holds. This proves our conclusion.

Example 8 By connecting the vertices of a regular n-gon we get a closed n-line. Prove that if n is even then among the connecting lines there are two parallel lines; if n is odd then it is impossible that there are only two parallel lines among the connecting lines.

Proof It is not suitable to solve this geometric problem by geometric methods. It concerns a complete system of residues. We mark these vertices anti-clockwise with numbers $0, 1, \ldots, n-1$. Assume that the closed n-line is $a_0 \to a_1 \to \cdots \to a_{n-1} \to a_n = a_0$, where $a_0, a_1, \ldots, a_{n-1}$ is a permutation of $0, 1, \ldots, n-1$.

At first, since all a_i are the vertices of the regular n-gon we get

$$a_i a_{i+1} \text{ // } a_j a_{j+1} \Leftrightarrow \widehat{a_{i+1} a_i} = \widehat{a_{j+1} a_j}$$
$$\Leftrightarrow a_i + a_{i+1} \equiv a_j + a_{j+1} \pmod{n}.$$

When n is even, $2 \nmid (n-1)$, thus the sum of numbers in any complete system of residues $\equiv 0 + 1 + \cdots + (n-1) = \dfrac{n(n-1)}{2} \not\equiv 0 \pmod{n}$.

But, on the other hand, we always have

$$\sum_{i=0}^{n-1} (a_i + a_{i+1}) = \sum_{i=0}^{n-1} a_i + \sum_{i=0}^{n-1} a_{i+1} = 2 \sum_{i=0}^{n-1} a_i = 2 \times \frac{n(n-1)}{2}$$
$$= n(n-1) \equiv 0 \pmod{n}. \qquad (6.11)$$

Hence, $a_i + a_{i+1} (i = 0, 1, \ldots, n-1)$ cannot form a complete system of residues modulo n, i.e., there must exist i and j $(0 \leqslant i, j \leqslant n-1)$, such that

$$a_i + a_{i+1} \equiv a_j + a_{j+1} \pmod{n}.$$

Thus there is a pair of sides $a_i a_{i+1}$ // $a_j a_{j+1}$.

When n is odd, if there is only one pair of sides $a_i a_{i+1}$ // $a_j a_{j+1}$, then among n numbers $a_0 + a_1$, $a_1 + a_2$, $\ldots$, $a_{n-1} + a_0$ there is one residue class r occurring twice. So it is short of one residue class s, thus (in this case $2 \mid (n-1)$)

$$\sum_{i=0}^{n-1} (a_i + a_{i+1}) \equiv 0 + 1 + \cdots + (n-1) + r - s$$
$$= \frac{n(n-1)}{2} + r - s$$
$$\equiv r - s \pmod{n}.$$

Combining with (6.11) we have $r \equiv s \pmod{n}$, a contradiction! This means that when n is odd it is impossible that there is only one pair of parallel sides.

Example 9   Let $n > 3$ be an odd number. Prove that after taking out arbitrarily one element from n-element set $S = \{0, 1, \ldots, n-1\}$ we can always classify the rest of the elements into two groups, every group consists of $\dfrac{n-1}{2}$ numbers, and the sums of the numbers in two groups are congruent modulo n.

Proof The first key in proof is for any $x \in S$, $x \neq 0$, the set $S\backslash\{x\}$ can be obtained from $T = \{1, 2, \ldots, n-1\}$ by the transformation

$$T + x \ (\mathrm{mod}\ n) = \{a + x \ (\mathrm{mod}\ n), a \in T\}.$$

This reduces the problem to prove the special situation: $T = S\backslash\{0\}$ can be classified into two groups, such that every group consists of $\dfrac{n-1}{2}$ numbers, and sums of elements in each group are congruent modulo n.

We divide into two cases. When $n = 4k + 1$ $(k \geqslant 1)$, note that in $2k$ pairs of numbers

$$\{1, 4k\}, \{2, 4k-1\}, \ldots, \{2k, 2k+1\},$$

sums of every pair are 0 modulo n, thus choose any k pairs of numbers as a set, the rest k pairs as another set, they satisfy the requirement.

When $n = 4k + 3$ $(k \geqslant 1)$, we first choose 1, 2, $4k$ as a set, 3, $4k+1$, $4k+2$ as another set. Then from the rest $2k-2$ pairs of numbers

$$\{4, 4k-1\}, \ldots, \{2k+1, 2k+2\}$$

choose $k-1$ pairs, respectively, and put them into the above two sets. The resulting sets are required sets.

Exercises

6.1 The vertices of a cube are labelled $+1$ or -1. Each face is assigned a number which is equal to the product of the numbers on vertices of the face. Prove that the sum of 14 assigned numbers is not 0.

6.2 Find all positive integers n such that the integers base 10 consisting of $n-1$ "1"s and one 7 are all prime.

6.3   Let p be a prime, $a \geqslant 2$, $m \geqslant 1$, $a^m \equiv 1 \pmod{p}$, $a^{p-1} \equiv 1 \pmod{p^2}$. Prove that $a^m \equiv 1 \pmod{p^2}$.

6.4   Let m be a positive integer. Prove that among the first m^2 terms of the sequence defined by

$$x_1 = x_2 = 1, \ x_{k+2} = x_{k+1} + x_k \, (k = 1, 2, \ldots)$$

there is a term which is divisible by m.

Some Famous Theorems in Number Theory

Fermat's little theorem, Euler's theorem and the Chinese remainder theorem, which are famous theorems of number theory, play an important role in elementary number theory.

(1) **Fermat's little theorem.** Let p be a prime, a any integer which is relatively prime to p. Then

$$a^{p-1} \equiv 1 \pmod{p}.$$

Fermat's little theorem possesses a variation which sometimes is more useful.

For any integer a, $a^p \equiv a \pmod{p}$. (When $p \nmid a$ these two statements are equivalent; when $p \mid a$, the later clearly holds.)

By induction it is not difficult to give a proof of Fermat's little theorem. It is easy to see that we need only to prove the statement for $a = 0, 1, \ldots, p - 1$. When $a = 0$ the conclusion is clearly true. If $a^p \equiv a \pmod{p}$ is true, then since $p \mid \binom{p}{i}$ $(i = 1, 2, \ldots, p - 1)$,

$$(a + 1)^p = a^p + \binom{p}{1}a^{p-1} + \cdots + \binom{p}{p-1}a + 1$$

$$\equiv a^p + 1 \equiv a + 1 \pmod{p},$$

it means that if we replace a by $a + 1$ the statement is also true.

(2) **Euler's theorem.** Let $m > 1$ be an integer, a an integer relatively prime to m, $\varphi(m)$ the Euler function (cf. Chapter 6), then

$$a^{\varphi(m)} \equiv 1 \pmod{m}.$$

We can prove Euler's theorem as follows. Put $r_1, r_2, \ldots, r_{\varphi(m)}$ a reduced system modulo m. Since $\gcd(a, m) = 1$, $ar_1, ar_2, \ldots,$

$ar_{\varphi(m)}$ is also a reduced system modulo m (cf. Chapter 6). Since two complete (reduced) systems modulo m are the same up to a permutation (under modulo m), in particular, we have

$$r_1 r_2 \cdots r_{\varphi(m)} \equiv ar_1 \cdot ar_2 \cdots ar_{\varphi(m)} = a^{\varphi(m)} r_1 r_2 \cdots r_{\varphi(m)} \pmod{m}.$$

Since $\gcd(r_i, m) = 1$, we have $\gcd(r_1 r_2 \cdots r_{\varphi(m)}, m) = 1$, thus from the above equation we can delete $r_1 \cdots r_{\varphi(m)}$, and have $a^{\varphi(m)} \equiv 1 \pmod{m}$.

Remark 1   When $m = p$ is a prime, since $\varphi(p) = p - 1$, from Euler's theorem we can get Fermat's little theorem.

Remark 2   If the standard factorization of m is given: $m = p_1^{a_1} \cdots p_k^{a_k}$, then the Euler function $\varphi(m)$ is determined by the following formula (the proof is omitted):

$$\varphi(m) = p_1^{a_1-1}(p_1 - 1) p_2^{a_2-1}(p_2 - 1) \cdots p_k^{a_k-1}(p_k - 1)$$

$$= m\left(1 - \frac{1}{p_1}\right)\left(1 - \frac{1}{p_2}\right) \cdots \left(1 - \frac{1}{p_k}\right).$$

(3) **The Chinese remainder theorem.** Let $m_1, m_2, \ldots, m_k$ be k pairwise relatively prime positive integers, $M = m_1 m_2 \cdots m_k$, $M_i = \dfrac{M}{m_i}$ ($i = 1, 2, \ldots, k$), $b_1, b_2, \ldots, b_k$ any integers. Then the system of congruences

$$x \equiv b_1 \pmod{m_1}, \ldots, x \equiv b_k \pmod{m_k}$$

has a unique solution

$$x \equiv M_1^* M_1 b_1 + M_2^* M_2 b_2 + \cdots + M_k^* M_k b_k \pmod{M},$$

where M_i^* are any integers such that

$$M_i^* M_i \equiv 1 \pmod{m_i} \ (i = 1, 2, \ldots, k).$$

To verify the above conclusion is an easy task, we leave it to readers (note that for any i we have $m_i | M$, and $m_i | M_j$ for any $j \neq i$). The main power of the Chinese remainder theorem is that it claims that there must exist a solution for the mentioned system of congruences when the modulo are pairwise relatively prime, and the precise form of the solution is usually not important.

The above-mentioned theorems of number theory are powerful tools in problem solving. They are used usually together with other methods. We will find it so latter. Now we only introduce some examples in which we directly apply these theorems.

Example 1　Let p be a given prime. Prove that in sequence $\{2^n - n\}$ $(n \geq 1)$ there are infinitely many terms which are divisible by p.

Proof　When $p = 2$ the statement is clearly true. If $p > 2$, then due to Fermat's little theorem, $2^{p-1} \equiv 1 \pmod{p}$, so for any positive integer m, we have

$$2^{m(p-1)} \equiv 1 \pmod{p}. \tag{7.1}$$

Put $m \equiv -1 \pmod{p}$, then by (7.1) we get

$$2^{m(p-1)} - m(p-1) \equiv 1 + m \equiv 0 \pmod{p}.$$

Hence, if $n = (kp - 1)(p - 1)$, then $2^n - n$ is divisible by p (for any positive integer k), thus in the sequence there are infinitely many terms which are divisible by p.

Example 2　Prove that in sequence 1, 31, 331, 3331, ... there are infinitely many composite numbers.

Proof　Since 31 is a prime, by Fermat's little theorem, $10^{30} \equiv 1 \pmod{31}$. Thus for any positive integer k we have $10^{30k} \equiv 1 \pmod{31}$. Hence

$$\frac{1}{3}(10^{30k} - 1) \equiv 0 \pmod{31}.$$

This means that the number consisting of $30k$ many 3's is divisible by 31. Multiply this number by 100 and then add 31, the resulting number is also divisible by 31, i. e., the $30k + 2$-th term in the sequence is divisible by 31, so it is not prime. Thus in the above sequence there are infinitely many composite numbers.

Example 3　Show that for any given positive number n, there are n consecutive positive integers such that every such positive integer has a square divisor greater than 1.

Proof　Since there are infinitely many primes, we can take out n

different primes p_1, p_2, ..., p_n. Consider the following system of congruences

$$x \equiv -i \pmod{p_i^2}, \quad i = 1, 2, \ldots, n. \tag{7.2}$$

Since p_1^2, p_2^2, ..., p_n^2 are pairwise relatively prime, by the Chinese remainder theorem we know that the above system of congruences has a positive integral solution. Thus n consecutive integers $x + 1$, $x + 2$, ..., $x + n$ are divisible by squares p_1^2, p_2^2, ..., p_n^2, respectively.

If we do not use primes directly, we can also adopt the following variation. Since Fermat's numbers $F_k = 2^{2^k} + 1$ $(k \geqslant 0)$ are pairwise relatively prime (cf. Example 3 in Chapter 2), after replacing p_i^2 in (7.2) by F_i^2 $(i = 1, 2, \ldots, n)$, the corresponding system of congruences has also a solution, which deduces the same result.

Remark The solution of Example 3 shows a basic function of the Chinese remainder theorem. It reduces the problem of "finding out n consecutive integers possessing some property" into "finding out n pairwise relatively prime numbers possessing some property", and the latter is easier to solve.

The following Example 4 contains some interesting skills.

Example 4 For any given positive integer n, there are n consecutive positive integers such that all such numbers are not power numbers (for the definition of power number, refer to Example 9 in Chapter 5).

Proof We will prove that there exist n consecutive positive integers, among them every number has at least one prime divisor which occurs just once in its unique prime factorization, thus it is not a power number. For this we choose n different primes p_1, p_2, ..., p_n, and consider the system of congruences

$$x \equiv -i + p_i \pmod{p_i^2}, \quad i = 1, 2, \ldots, n. \tag{7.3}$$

Since p_1^2, p_2^2, ..., p_n^2 are pairwise relatively prime, by the Chinese remainder theorem the above system of congruences possesses a positive integral solution. Since $x + i \equiv p_i \pmod{p_i^2}$ for $1 \leqslant i \leqslant n$, $p_i \mid (x + i)$. But by (7.3) we know $p_i^2 \nmid (x + i)$, i.e., p_i occurs just once in the unique prime factorization of $x + i$, thus none of $x + 1$, $x + 2, \ldots, x + n$ is a power number.

Example 5 For a given positive integer n, let $f(n)$ be the minimal positive integer such that $\sum_{k=1}^{f(n)} k$ is divisible by n. Prove that $f(n) = 2n - 1$ if and only if n is a power of 2.

Proof The first part of the problem is quite easy. If $n = 2^m$, then on the one hand,

$$\sum_{k=1}^{2n-1} k = \frac{(2n - 1) \times 2n}{2} = (2^{m+1} - 1) \cdot 2^m$$

is divisible by $2^m = n$. On the other hand, if $r \leqslant 2n - 2$, then

$$\sum_{k=1}^{r} k = \frac{r(r + 1)}{2}$$

is not divisible by 2^m since one of r and $r + 1$ is odd and the other one is not great than $(2n - 2) + 1 = 2^{m+1} - 1$, thus is not divisible by 2^{m+1}. Combining these two parts, we know $f(2^m) = 2^{m+1} - 1$.

Now assume that n is not a power of 2, i.e., $n = 2^m a$, where $a > 1$ is odd. We want to show that there exists a positive integer $r < 2n - 1$, such that $2^{m+1} \mid r$ and $a \mid (r + 1)$, thus

$$\sum_{k=1}^{r} k = \frac{r(r + 1)}{2}$$

is divisible by $2^m a = n$, so $f(n) < 2n - 1$.

To prove the above assertion we consider

$$x \equiv 0 \pmod{2^{m+1}}, \quad x \equiv -1 \pmod{a}. \tag{7.4}$$

Since $\gcd(2^{m+1}, a) = 1$, by the Chinese remainder theorem there is a solution x_0 for the system of congruences (7.4) and all its solutions are $x \equiv x_0 \pmod{2^{m+1}a}$, i.e., $x \equiv x_0 \pmod{2n}$. Hence we can determine a solution r satisfying (7.4) and $0 < r \leqslant 2n$. Furthermore, by the second congruence in (7.4) we know $r \neq 2n$. However from the first congruence we get $r \neq 2n - 1$. So $r < 2n - 1$. This shows the existence of r meeting the requirement.

Example 6   Let n and k be given integers, $n > 0$ and $k(n - 1)$ be even. Prove that there are x and y such that $\gcd(x, n) = \gcd(y, n) = 1$ and $x + y \equiv k \pmod{n}$.

Proof   We first show that when n is a power of prime p^a the conclusion is true. Actually, we can prove that there are x and y such that $p \nmid xy$ and $x + y = k$ as follows:

If $p = 2$ then the condition shows that k is even. In this case we can take $x = 1$, and $y = k - 1$; if $p > 2$ then one of pairs $x = 1$, $y = k - 1$ or $x = 2$, $y = k - 2$ meets the requirement.

In general case, let $n = p_1^{q_1} p_2^{q_2} \cdots p_r^{q_r}$ be the unique standard factorization of n. We have proved above that for every p_i there are integers x_i and y_i such that $p_i \nmid x_i y_i$ and $x_i + y_i = k$ $(i = 1, 2, \ldots, r)$. But by the Chinese remainder theorem the following system of congruences

$$x \equiv x_i \pmod{p_i^{q_i}} (i = 1, 2, \ldots, r) \tag{7.5}$$

has a solution x, and the system of congruences

$$y \equiv y_i \pmod{p_i^{q_i}} (i = 1, 2, \ldots, r) \tag{7.6}$$

has a solution y. Now it is not difficult to verify that solutions x and y meet the requirements of the problem for $p_i \nmid x_i y_i$, so $p_i \nmid xy$ $(i = 1, 2, \ldots, r)$, thus $\gcd(xy, n) = 1$. By (7.5) and (7.6), we have

$$x + y \equiv x_i + y_i = k \pmod{p_i^{q_i}} (i = 1, 2, \ldots, r).$$

So $x + y \equiv k \pmod{n}$.

Remark   The proof of Example 6 shows the basic role of the Chinese remainder theorem: we can reduce a problem about an arbitrary positive integer n to a problem about the power of prime n, and the latter one is usually easier to solve.

Exercises

7. 1   Let n be an odd prime, and $n = \dfrac{2^{2p} - 1}{3}$. Prove that $2^{n-1} \equiv 1$ (mod n).

7. 2   Let x be a given positive integer. Prove that there are n consecutive positive integers such that each of them is not a power of a prime.

7. 3   Let m and n be positive integers possessing the following property: the equation

$$\gcd(11k - 1, m) = \gcd(11k - 1, n)$$

holds for all positive integers k. Prove that $m = 11^r n$ for some integer r.

8   Order and its Application

Let $n > 1$ and a be integers with gcd $(a, n) = 1$. Then there is r $(1 \leqslant r \leqslant n - 1)$ such that $a^r \equiv 1 \pmod{n}$.

In fact, since n numbers $a^0, a^1, \ldots, a^{n-1}$ are all relatively prime to n, after modulo n they have at most $n - 1$ different remainders, so among them there are two numbers which are congruent modulo n. That is, there are $0 \leqslant i < j \leqslant n - 1$, such that $a^i \equiv a^j \pmod{n}$, and $a^{j-i} \equiv 1 \pmod{n}$. Thus $r = j - i$ meets the requirement.

The minimal positive integer r satisfying $a^r \equiv 1 \pmod{n}$ is called the order of a modulo n. By the discussion above we know $1 \leqslant r \leqslant n - 1$. The following (1) shows that the order of a modulo n possesses a very sharp property:

(1) Assume that gcd $(a, n) = 1$ and the order of a modulo n is r. If there is a positive integer N such that $a^N \equiv 1 \pmod{n}$, then $r \mid N$.

This is true, because we can assume $N = rq + k$ $(0 \leqslant k < r)$, then

$$1 \equiv a^N \equiv (a^r)^q \cdot a^k \equiv a^k \pmod{n}.$$

As $0 \leqslant k < r$, by the above equation and the definition of r we have $k = 0$. Thus $r \mid N$.

By property (1) together with the Euler's theorem ((2) in Chapter 7) we have:

(2) Assume gcd $(a, n) = 1$, then the order of a modulo n divides $\varphi(n)$. In particular, if n is a prime p, then the order of a modulo p divides $p - 1$.

In many problems finding out the order of a modulo n is often very important. By using the order of a modulo n and property (1), from indexes of some powers of integers we can obtain some divisibility relations, this is a basic method deducing divisibility relations. On the other hand, determination of the order of a modulo n is usually very difficult. We can do it only in some special cases. For a specific a and n, by calculating the remainders of a, a^2, ... modulo n one by one we can determine the order of a modulo n. If we apply (2), this procedure can be simplified slightly.

Order is a powerful tool in solving many problems. We give some examples below.

Example 1 Assume $n > 1$, and $n \mid (2^n + 1)$, show that $3 \mid n$.

Proof Clearly n is odd. Let p be the minimum prime divisor of n. We will prove $p = 3$, which implies $3 \mid n$.

Let r be the order of 2 modulo p. By $2^n \equiv -1 \pmod{n}$ we have

$$2^{2n} \equiv 1 \pmod{p}. \qquad (8.1)$$

As $p \geqslant 3$, by Fermat's little theorem, we get

$$2^{p-1} \equiv 1 \pmod{p}. \qquad (8.2)$$

(8.1), (8.2) and the properties of order imply $r \mid 2n$ and $r \mid (p-1)$, so $r \mid \gcd(2n, p-1)$. It is easy to prove $\gcd(2n, p-1) = 2$. This is because from $2 \nmid n$ one has $2 \mid \gcd(2n, p-1)$, and $2^2 \nmid \gcd(2n, p-1)$. On the other hand, if there is an odd prime q such that $q \mid \gcd(2n, p-1)$, then $q \mid (p-1)$ and $q \mid n$, but the former shows $q < p$, this contradicts the fact that p is the minimal prime divisor of n. So $\gcd(2n, p-1) = 2$, thus $r = 2$, and $p = 3$.

The key idea of Example 1 is to consider (the minimum) prime divisor p of n, and modulo p to deduce the result.

Example 2 Let $n > 1$. Prove $n \nmid (2^n - 1)$.

Proof One We prove it by contradiction. Suppose that there is $n > 1$ such that $n \mid (2^n - 1)$. For any prime divisor p of n, one has $p \geqslant 3$. Assume that the order of 2 modulo p is r, then clearly $r > 1$. From $2^n \equiv 1 \pmod{n}$ one has

$$2^n \equiv 1 \pmod{p}. \qquad (8.3)$$

By Fermat's little theorem we get

$$2^{p-1} \equiv 1 \ (\text{mold} \ p). \tag{8.4}$$

Hence $r \mid n$ and $r \mid (p-1)$. Thus $r \mid \gcd (n, p-1)$. In particular, we take p being the minimum prime divisor of n, then $\gcd (n, p-1) = 1$. This is because if there is a prime q such that $q \mid \gcd (n, p-1)$, then $q \mid (p-1)$ and $q \mid n$. But the former means $q < p$, this contradicts the choice of p. Hence $\gcd (n, p-1) = 1$. So $r = 1$, a contradiction.

Remark 1 The key of this solution is the consideration of prime divisors of n. As $n > 1$ is equivalent to n having a prime divisor, we proceed from $2^n \equiv 1 \ (\text{mod} \ n)$ to congruence (8.3), although the assumption is weakened, it still depicts $n > 1$.

Congruence modulo a prime number usually gives more suitable properties (or results). For the purpose of this problem, the benefit of doing this is to obtain congruence (8.4). Example 1 and the following Example 3 are such cases.

Remark 2 Congruences (8.3) and (8.4) hold for any prime divisor p of n. Hence at the beginning of Proof One we consider p as a parameter to be determined, and deduce $r \mid \gcd(p-1, n)$, and supply the chance of choosing p to produce a contradiction.

Keeping the parameters unchanged provides us with more choices and maintain some flexibility. This is a basic method.

Remark 3 By Example 10 in Chapter 5 there are infinitely many n satisfying the condition in Example 1, it is exactly the opposite conclusion of Example 2. Readers can check carefully what difference in proof deduces such different conclusions.

Remark 4 By the way, if we do not use orders we can also solve Example 2. Let p be the minimum prime divisor of n, thus $\gcd (p-1, n) = 1$. But according to (8.3) and (8.4), one has $p \mid \gcd(2^{p-1} - 1, 2^n - 1)$, so Example 4 in Chapter 2 deduces $p \mid (2^{\gcd(p-1, n)} - 1)$, hence $p \mid 1$, we get a contradiction.

Example 1 can be proved similarly.

Proof Two This solution does not need to consider prime divisors of n. If there is $n > 1$ such that $n \mid \gcd(2^n - 1)$, then n is odd. Let r be the order of 2 modulo n. Then by $2^n \equiv 1 \pmod{n}$, we have $r \mid n$. But $2^r \equiv 1 \pmod{n}$. Therefore $2^r \equiv 1 \pmod{r}$, i.e.,

$$r \mid (2^r - 1). \tag{8.5}$$

Since the order r satisfies $1 \leqslant r < n$, and clearly $r \neq 1$ (otherwise we deduce $n = 1$), thus $1 < r < n$. By (8.5) we repeat the above argument, and get infinitely many integers $r_i (i = 1, 2, \ldots)$ such that $r_i \mid 2^{r_i} - 1$ and $n > r > r_1 > r_2 > \cdots > 1$, which is impossible.

In this proof we can use a simpler expression: take $n > 1$ to be the minimum integer such that $n \mid (2^n - 1)$, the above argument produces an integer r, such that $r \mid (2^r - 1)$ and $1 < r < n$, and it contradicts the choice of n.

Remark The method in Proof Two is the so-called infinitely decreasing method. Its basic idea is: by contradiction assume that there is a solution, we try to create another positive integer solution, and the new solution is "strictly smaller" than the original one, i.e., strictly decreasing. If the above process can be done an infinite number of times, then as strictly decreasing positive integer sequence has only finite terms, it leads to a contradiction.

Example 3 Let $n > 1$, $2 \nmid n$. Then for any integer m, $n \nmid (m^{n-1} + 1)$.

Proof Assume that there is an odd n great than 1 such that $n \mid (m^{n-1} + 1)$, then $\gcd(m, n) = 1$. Let p be any prime divisor of n, r is the order of m modulo p (note that $p \nmid m$). Also let $n - 1 = 2^k t$, $k \geqslant 1$, $2 \nmid t$, then we have

$$m^{2^k t} \equiv -1 \pmod{p}. \tag{8.6}$$

Thus $m^{2^{k+1} t} \equiv 1 \pmod{p}$, so $r \mid 2^{k+1} t$.

The key point is to prove $2^{k+1} \mid r$. Suppose that the result is not true. Then by $m^r \equiv 1 \pmod{p}$ one gets $m^{2^k t} \equiv 1 \pmod{p}$, together with (8.6) we have $p = 2$, a contradiction. Therefore $2^{k+1} \mid r$.

Now $\gcd(p, m) = 1$ implies $m^{p-1} \equiv 1 \pmod{p}$, thus $r \mid (p - 1)$, so $2^{k+1} \mid (p - 1)$, i.e., $p \equiv 1 \pmod{2^{k+1}}$. As p is any prime divisor of n, we can factorize n and get $n \equiv 1 \pmod{2^{k+1}}$, i.e., $2^{k+1} \mid (n - 1)$.

But this contradicts the above assumption $2^k \parallel (n-1)$.

Example 4   Let p be an odd prime. Prove that any positive divisor of $\dfrac{p^{2p}+1}{p^2+1}$ is congruent to 1 modulo $4p$.

Proof   It is sufficient to prove that any prime divisor of $\dfrac{p^{2p}+1}{p^2+1}$ satisfies $q \equiv 1 \pmod{4p}$. First, we note that

$$\frac{p^{2p}+1}{p^2+1} = p^{(p-1)} - p^{2(p-2)} + \cdots - p^2 + 1. \tag{8.7}$$

Hence $q \neq p$. Let r be the order of p modulo q. As

$$p^{2p} \equiv -1 \pmod{q}, \tag{8.8}$$

so $p^{4p} \equiv 1 \pmod{q}$, therefore $r \mid 4p$. Thus $r = 1, 2, 4, p, 2p$ or $4p$.

If $r = 1, 2, p, 2p$, then $p^{2p} \equiv 1 \pmod{q}$, together with (8.8) we have $q = 2$. This is impossible. If $r = 4$, then q is a prime, which implies $q \mid (p^2 - 1)$ or $q \mid (p^2 + 1)$. The former is proved to be impossible. If the latter holds, i.e., $p^2 \equiv -1 \pmod{q}$. We consider (8.7) modulo q. Of course the left side modulo q is 0. But the right side $\equiv (-1)^{p-1} - (-1)^{p-2} + \cdots - (-1) + 1 \equiv p \pmod{q}$. Hence $p = q$, which is impossible. Thus $r \neq 4$. Hence $r = 4p$.

Finally, as $\gcd(p, q) = 1$, by Fermat's little theorem $p^{q-1} \equiv 1 \pmod{q}$. Thus $r \mid (q-1)$, i.e., $4p \mid (q-1)$, hence $q \equiv 1 \pmod{4p}$.

In the above solution, the key point is to determine the order of p modulo q. The following Example 5 is an interesting result about orders, its proof has also a certain degree of universality.

Example 5   (1) Let p be an odd prime, $q \neq \pm 1$, and $p \nmid a$. Assume that r is the order of a modulo p, and k_0 satisfies $p^{k_0} \parallel (a^r - 1)$. Denote the order of a modulo p^k by r_k, then

$$r_k = \begin{cases} r, & \text{if } k = 1, \ldots, k_0, \\ rp^{k-k_0}, & \text{if } k > k_0. \end{cases}$$

(2) Assume that a is odd, $a \equiv 1 \pmod{4}$, $a \neq 1$, k_0 satisfies $2^{k_0} \parallel (a-1)$. Denote the order of a modulo 2^k by l_k, then

$$l_k = \begin{cases} 1, & \text{if } k = 1, \ldots, k_0, \\ 2^{k-k_0}, & \text{if } k > k_0. \end{cases}$$

(3) Assume that a is odd, $a \equiv -1 \pmod 4$, $a \neq -1$, k_0 satisfies $2^{k_0} \parallel (a+1)$. Denote the order of a modulo 2^k by l_k, then

$$l_k = \begin{cases} 1, & \text{if } k = 1, \\ 2, & \text{if } k = 2, \ldots, k_0 + 1, \\ 2^{k-k_0}, & \text{if } k > k_0 + 1. \end{cases}$$

Proof (1) When $1 \leqslant k \leqslant k_0$, $a^{r_k} \equiv 1 \pmod{p^k}$ implies $a^{r_k} \equiv 1 \pmod p$. According to the definition of r we have $r \mid r_k$. On the other hand, $a^r \equiv 1 \pmod{p^{k_0}}$ implies $a^r \equiv 1 \pmod{p^k}$. Thus from the definition of r_k we get $r_k \mid r$. Thus $r_k = r$ $(k = 1, \ldots, k_0)$.

Now let $k > k_0$. At first, we prove that for every $i = 0, 1, \ldots$, $p^{k_0+i} \parallel (a^{rp^i} - 1)$, that is,

$$a^{rp^i} = 1 + p^{k_0+i}u_i, \quad \gcd(u_i, p) = 1. \tag{8.9}$$

It can be proved by induction. When $i = 0$, according to the definition of k_0 we have that (8.9) is true. Assume that (8.9) is true for $i \geqslant 0$, then by the binomial theorem,

$$a^{rp^{i+1}} = (1 + p^{k_0+i}u_i)^p = 1 + p^{k_0+i+1}u_i + \binom{p}{2}p^{2k_0+2i}u_i^2 + \cdots$$

$$= 1 + p^{k_0+i+1}\left(u_i + \binom{p}{2}p^{k_0+i-1}u_i^2 + \cdots\right)$$

$$= 1 + p^{k_0+i+1}u_{i+1},$$

where $p \nmid u_{i+1}$ (note that we need the condition $p \geqslant 3$), thus (8.9) is true for all $i \geqslant 0$.

By using (8.9), for $k \geqslant k_0$ we will prove $r_k = rp^{k-k_0}$ by induction. When $k = k_0$, the above argument has proved the result. If $k > k_0$, assume that $r_{k-1} = rp^{k-k_0-1}$ is true. On one hand, in (8.9) we take $i = k - k_0$ then $a^{rp^{k-k_0}} \equiv 1 \pmod{p^k}$, thus $r_k \mid rp^{k-k_0}$. On the other hand, from $a^{r_k} \equiv 1 \pmod{p^k}$ we have $a^{r_k} \equiv 1 \pmod{p^{k-1}}$, so $r_{k-1} \mid r_k$, thus $r_k = rp^{k-k_0}$ or rp^{k-k_0-1}. But in (8.9) put $i = k - k_0 - 1$, we get $a^{rp^{k-k_0-1}} \not\equiv \mid \pmod{p^k}$, so $r_k = rp^{k-k_0}$.

(2) When $1 \leqslant k \leqslant k_0$, the result is clearly true. When $k > k_0$, note that $a \equiv 1 \pmod 4$, $a \neq 1$ implies $k_0 \geqslant 2$. According to the above information by induction about $i = 0, 1, \ldots$, it is easy to prove

$$a^{2^i} = 1 + 2^{k_0+i} u_i, \quad 2 \nmid u_i. \tag{8.10}$$

By (8.10), with the same proof as (1) it is not difficult to get

$$l_k = 2^{k-k_0} \ (k \geqslant k_0).$$

(3) By $a \equiv -1 \pmod 4$, it is easy to prove that the result is true when $k = 1, 2, \ldots, k_0 + 1$. By induction, for $i = 1, 2, \ldots$, we have

$$a^{2^i} = 1 + 2^{k_0+i} u_i, \quad 2 \nmid u_i. \tag{8.11}$$

By the same proof as (1) we can get $l_k = 2^{k-k_0}$ (for $k \geqslant k_0 + 1$).

Remark 1 Let a and $n > 0$ be two given pairwise relatively prime integers, and not ± 1. Assume that $n = 2^a p_1^{q_1} \cdots p_k^{q_k}$ (p_i are odd primes, $a \geqslant 0$) is the standard factorization of n. If the order of a modulo 2^a is determined, then by Example 5 one can determine the order of a modulo $p_i^{q_i}$ and the order of a modulo 2^a, too. Furthermore, according to the result in Exercise 8.2, the order of a modulo n can be obtained. Hence, in order to determine the order of a modulo an integer n, eventually it is reduced to determining the order of a modulo an odd prime p. Generally speaking, the latter is a very difficult problem, but for small a and p, we can calculate it by hand.

Remark 2 Assume that p is an odd prime, $a \neq \pm 1$, $p \nmid a$, r is the order of a modulo p, and k_0 satisfies $p^{k_0} \parallel (a^r - 1)$. Then from the proof of (8.9) in Example 5, for any positive integer m relatively prime to p we have

$$a^{mrp^i} = 1 + p^{k_0+i} u_i', \quad \gcd(u_i', p) = 1, \ i = 0, 1, \ldots.$$

According to this result and notice that p and r must be relatively prime, we can easily get the result below.

(1) Assume that a positive integer n satisfies $r \mid n$ and $p^l \parallel n$, then

$$p^l \left\| \frac{a^n - 1}{a^r - 1} \right..$$

Moreover, assume that a is odd, $a \neq \pm 1$, k_0 satisfies $2^{k_0} \parallel (a^2 - 1)$, m is any odd positive number, then

$$a^{2^i m} = 1 + 2^{k_0 + i - 1} u_i', \quad 2 \nmid u_i', \quad i = 1, 2, \ldots.$$

Thus we find that the following holds.

(2) Let n be a positive integer, $2^l \parallel n$. If $l \geqslant 1$, then $2^{l-1} \left\| \dfrac{a^n - 1}{a^2 - 1} \right.$.

The above forms of (8.9), (8.10) and (8.11) in Example 5 sometimes are more convenient to use.

Remark 3 Let p be an odd prime, a and b integers, $p \nmid ab$, then there is a positive integer r, such that

$$a^r \equiv b^r (\text{mod } p). \qquad (8.12)$$

This is because there is b_1 such that $bb_1 \equiv 1 \,(\text{mod } p)$, and also there is a positive integer r satisfying $(ab_1)^r \equiv 1 \,(\text{mod } p)$, which deduces (8.12). Moreover, it is easy to see that the minimun positive integer r such that (8.12) holds is equal to the order of ab_1 modulo p. Hence, if a positive integer n satisfies $a^n \equiv b^n \,(\text{mod } p)$, then $r \mid n$.

(1) and (2) in Remark 2 has the following generalization. Its proof is similar.

(1) Assume $a \neq \pm b$, and n is a positive integer. If $r \mid n$, $p^l \parallel n$, then $p^l \left\| \dfrac{a^n - b^n}{a^r - b^r} \right.$.

(2) Assume that a and b are odd numbers, $a \neq \pm b$, n is a positive integer, $2^l \parallel n$. If $l \geqslant 1$, then $2^{l-1} \left\| \dfrac{a^n - b^n}{a^2 - b^2} \right.$.

Example 6 Suppose that a and n are integers, none of them equals ± 1, and $\gcd(a, n) = 1$. Prove that there are at most finitely many k such that $n^k \mid (a^k - 1)$.

Proof As $n \neq \pm 1$, n has prime divisors. Firstly we assume that n has an odd prime divisor p, then $p \nmid a$. Let r be the order of a modulo p. Since $a \neq \pm 1$, there is a positive integer k_0 such that $p^{k_0} \parallel (a^r - 1)$.

If there are infinitely many k such that $n^k \mid (a^k - 1)$, so there are infinitely many $k > k_0$, such that

$$a^k \equiv 1 \pmod{p^k}. \tag{8.13}$$

But according to Example 5, the order of a modulo p^k is rp^{k-k_0}. Hence by (8.13) we have $rp^{k-k_0} \mid k$, thus $k \geqslant rp^{k-k_0} \geqslant 3^{k-k_0}$, and the number of such k is clearly finite, a contradiction.

If n does not have odd prime divisors, then n is a power of 2. Firstly note that if odd k satisfies $n^k \mid (a^k - 1)$, then

$$a^k - 1 = (a - 1)(a^{k-1} + \cdots + a + 1) \tag{8.14}$$

is divisible by 2^k. But the second divisor in (8.14) is the sum of an odd number of odd integers, so $2^k \mid (a - 1)$. As $a \neq 1$, there is at most a finite number of such k.

Assume that there are infinitely many even numbers $k = 2l$ such that $n^k \mid (a^k - 1)$, then

$$(a^2)^l \equiv 1 \pmod{2^l}. \tag{8.15}$$

Define k_0 satisfying $2^{k_0} \parallel (a^2 - 1)$, then $k_0 \geqslant 3$. By Example 5, when $l > k_0$ the order of a^2 modulo 2^l is 2^{l-k_0}. Hence if $l > k_0$, by (8.15) we have $2^{l-k_0} \mid l$, thus $l \geqslant 2^{l-k_0}$. But there are at most finite number of such l, a contradiction!

Exercises

8.1 Prove that for any Fermat's number $F_k = 2^{2^k} + 1$ ($k \geqslant 0$) its divisors $\equiv 1 \pmod{2^{k+1}}$.

8.2 (1) Let m and n be relatively prime positive integers, m, $n > 1$, a an integer relatively prime to mn. Assume that d_1, d_2 are the orders of a modulo m and n, respectively, then the order of a modulo mn is $[d_1, d_2]$.

(2) Find the order of 3 modulo 10^4.

8.3 Prove that for any integer $k > 0$, there is a positive integer n, such that $2^k \mid (3^n + 5)$.

Indeterminate Equations (II)

Congruence is a powerful tool in solving indeterminate equations. We usually apply congruence to prove that indeterminate equations have no integer solutions, or deduce some necessary conditions the solutions satisfy. These proofs are often versatile. They occur frequently in mathematical competition. In this chapter we will choose some examples in this area to show the applications of congruence.

Example 1   If $n \equiv 4 \pmod 9$, prove that the determinate equation

$$x^3 + y^3 + z^3 = n \qquad (9.1)$$

has no integer solutions (x, y, z).

Proof   If equation (9.1) has integer solutions, then (9.1) modulo 9 has also integer solutions. It is well-known that a perfect cube modulo 9 is congruent to one of 0, 1, -1. Hence

$$x^3 + y^3 + z^3 \equiv 0, 1, 2, 3, 6, 7, 8 \pmod 9.$$

But $n \equiv 4 \pmod 9$, so (9.1) modulo 9 has no solution. This contradicts the above argument. Therefore, equation (9.1) has no integer solutions.

In dealing with indeterminate equations using congruence method, the key point is to choose a suitable modulo. Example 1 is a relatively easy problem as modulo 9 has occurred in the problem. In comparison, the following Example 2 is somewhat difficult.

Example 2   Find all nonnegative integer solutions $(x_1, \ldots, x_{14})$ (the order of integers in the solutions is immaterial) of the following equation

$$x_1^4 + x_2^4 + \cdots + x_{14}^4 = 1599.$$

Solution   Applying modulo 16 we can prove that the equation has no integer solutions, because the power of an integer modulo 16 is congruent to 0 or 1, all possible values of $x_1^4 + x_2^4 + \cdots + x_{14}^4$ modulo 16 are 0, 1, 2, $\ldots$, 14, not 15. But $1599 \equiv 15 \pmod{16}$. Thus the equation has no such solutions. The result is proved.

The reason of choosing 16 is that there are 14 terms on the left side of the equation and choosing the number of residue classes $\geqslant 15$ is relatively hopeful to get a contradiction (here we use congruence method to prove that there is no integer solutions for the equation). However $15 = 3 \times 5$, and according to Chinese remainder theorem modulo 15 is equivalent to modulo 3 and modulo 5, it does not work.

Example 3   Prove that the following numbers cannot be expressed in terms of the sum of the cubes of some consecutive integers.

(1) 385^{97};

(2) 366^{17}.

Proof   By using

$$1^3 + 2^3 + \cdots + k^3 = \left(\frac{k(k+1)}{2}\right)^2,$$

it is easy to obtain that the sum of cubes of some consecutive integers can be expressed in the form

$$\left(\frac{m(m+1)}{2}\right)^2 - \left(\frac{n(n+1)}{2}\right)^2. \tag{9.2}$$

It can be divided into two cases: the integers are all positive, and there are both positive and negative integers. We have to prove that for integers in (1) or (2), there do not exist m and n such that it can be expressed in the form (9.2). Although using decomposition method, in principle, the problem can be solved, however, it is considerably hard. Using congruence is quite straightforward.

At first, according to integer x modulo 9 we can classify them and test them one by one. We easily have that $\left(\frac{x(x+1)}{2}\right)^2$ modulo 9 is congruent to 0 or -1. Hence the numbers of the form (9.1) modulo 9 are 0, 1, -1. On the other hand, by Euler's theorem

$$385^{97} \equiv 385 \times (385^{16})^6 \equiv 385 \equiv 7 \ (\text{mod } 9),$$

proving the above statement.

However, as $366^{17} \equiv 0 \ (\text{mod } 9)$, for the number 366^{17}, modulo 9 does not work.

Now we choose modulo 7. It is easy to verify that for integer x, $\left(\dfrac{x(x+1)}{2}\right)^2$ modulo 7 is congruent to 0, 1, -1. Thus the numbers of the form (9.2) modulo 7 can only be 0, ± 1, ± 2. But

$$366^{17} \equiv 2^{17} \equiv 2 \times 2^4 \equiv 4 \ (\text{mod } 7).$$

Hence, our assertion is true.

For equations having integer solutions it is not easy to solve them by congruence. We need to use congruence together with other methods (such as estimation, factorization, etc.). Let us give some examples.

Example 4　Find all powers of 2, such that after deleting its first digit (in decimal system expression) the new number is also a power of 2.

Solution　The problem is to find all positive integer solutions (n, k, m, a) of the equation

$$2^n = 2^k + a \times 10^m, \tag{9.3}$$

where $a = 1, 2, \ldots, 9$. Rewrite formula (9.3) as

$$2^k(2^{n-k} - 1) = a \times 10^m. \tag{9.4}$$

First, we prove $m = 1$. Since if $m > 1$, then the right side of (9.4) is divisible by 5^2, so $5^2 \mid (2^{n-k} - 1)$. Moreover, it is easy to find that the order of 2 modulo 5^2 is 20 (note that the order divides $\varphi(25) = 20$, and $2^{10} \equiv -1 \ (\text{mod } 25)$). Hence 20 divides $n - k$, thus $2^{20} - 1$ divides the left side of (9.4). But $2^{20} - 1 = (2^5)^4 - 1$ has divisor $2^5 - 1 = 31$. However, 31 does not divide the right side of (9.3), a contradiction! Therefore $m = 1$.

Now we need only to check among 2-digit numbers which power

of 2 is a required solution, and these are only 32 and 64.

Example 5   Find all positive integers $x > 1$, $y > 1$ and $z > 1$, such that

$$1! + 2! + \cdots + x! = y^z. \tag{9.5}$$

Solution   The key step is to prove that $x \geqslant 8$ implies $z = 2$. Since the left side of (9.5) is divisible by 3, $3 \mid y^z$, hence $3 \mid y$. Thus the right side of (9.5) is divisible by 3^z. On the other hand,

$$1! + 2! + \cdots + 8! = 46\ 233$$

is divisible by 3^2, but is not divisible by 3^3. However, when $n \geqslant 9$, we have $3^3 \mid n!$. So when $x \geqslant 8$, the left side of (9.5) is divisible by 3^2 but not by 3^3. Thus so does the right side of (9.5), i.e., $z = 2$.

Furthermore, we prove that when $x \geqslant 8$ equation (9.5) has no solutions. When $x \geqslant 8$ the left side of (9.5) $\equiv 1! + 2! + 3! + 4! \equiv 3$ (mod 5). We have proved that in this case $z = 2$, so the right side of (9.5) $y^2 \equiv 0, \pm 1$ (mod 5), thus the above statement holds.

Finally, when $x < 8$, by checking it is easy to find that the solution of (9.5) is $x = y = 3$, $z = 2$.

In Example 4 and Example 5, by comparing indexes of powers of some prime occurring in the two sides of an equation, we deduce the results. This kind of method in congruence is called the method of comparing the powers of prime. In the following Example 6 we apply this method.

Example 6   Prove that the indeterminate equation

$$(x + 2)^{2m} = x^n + 2 \tag{9.6}$$

has no positive integer solutions.

Proof   For the following proof, we first deduce some simple conclusions from equation (9.6).

Clearly $n > 1$, and x is odd, otherwise, equation (9.6) modulo 4 will lead to a contradiction. Furthermore, n is odd, too, because if $2 \mid n$, then x^n is a square of an odd, so the right side of (9.6) $\equiv 1 + 2 = 3$ (mod 4), but the left side $\equiv 1$ (mod 4), it is impossible. Thus $2 \nmid n$.

Assume that $x + 1 = 2^a x_1$, where x_1 is odd, $a > 0$ (as x is odd). Rewrite equation (9.6) as follows

$$(x + 2)^{2m} - 1 = x^n + 1. \tag{9.7}$$

The left side of (9.7) has the divisor

$$(x + 2)^2 - 1 = (2^a x_1 + 1)^2 - 1 = 2^{a+1}(2^{a-1} x_1^2 + x_1),$$

so 2^{a+1} divides the left side of (9.7). But on the other hand, as $n - 1 > 0$ is even, by the binomial theorem we have

$$x^n + 1 = x(2^a x_1 - 1)^{n-1} + 1 \equiv x \cdot 1 + 1 = 2^a x_1 \pmod{2^{a+1}}.$$

Since $2 \nmid x_1$, the right of (9.7) $x^n + 1 \not\equiv 0 \pmod{2^{a+1}}$, a contradiction!

Example 7 Prove that all positive integer solutions of the equation

$$8^x + 15^y = 17^z \tag{9.8}$$

are $x = y = z = 2$.

Proof First, by congruence we prove that y and z are all even. Equation (9.8) modulo 4 is

$$(-1)^y \equiv 1 \pmod 4,$$

thus y is even. Equation (9.8) modulo 16 is

$$8^x + (-1)^y \equiv 1 \pmod{16},$$

that is, $8^x \equiv 0 \pmod{16}$, so $x \geq 2$.

Note that $17^2 \equiv 1$, $15^2 \equiv 1 \pmod{32}$, thus if z is odd, then from $2 \mid y$ and $x \geq 2$ together with (9.8) we have

$$1 \equiv 17 \pmod{32},$$

it is impossible. Therefore z is even.

Let $y = 2y_1$, $z = 2z_1$, then Equation (9.8) can be factorized into

$$(17^{z_1} - 15^{y_1})(17^{z_1} + 15^{y_1}) = 8^x. \tag{9.9}$$

It is easy to see that the greatest common divisor of the two divisors of the left side of (9.9) is 2, but the right side of (9.9) is a power of 2, so we get

$$17^{z_1} - 15^{y_1} = 2, \tag{9.10}$$
$$17^{z_1} + 15^{y_1} = 2^{3x-1}. \tag{9.11}$$

Equation (9.10) modulo 32 deduces that z_1 and y_1 must be odd (otherwise, the left side of (9.10)$\equiv 0$, -14, 16 (mod 32)). Adding (9.10) and (9.11), we have

$$17^{z_1} = 1 + 2^{3x-2}. \tag{9.12}$$

If $x \geqslant 3$, then the right side of (9.12)$\equiv 1$ (mod 32). But as z_1 is odd, the left side $\equiv 17$ (mod 32), it is impossible. Thus $x = 2$. By this together with (9.12) we have $z_1 = 1$, i.e., $z = 2$. Thus $y_1 = 1$, i.e., $y = 2$. Therefore $x = y = z = 2$.

This solution is a classic example of using congruence together with factorization. Using congruence we can deduce that y and z are all even. This is a preparation for the following factorization of equations.

The following Example 8 is more difficult. Here we introduce two solutions. In the first one we base on congruence together with factorization, which is quite simple. In the second one we use the method of comparing the powers of prime. Although it is trouble-some, it is typical.

Example 8 Prove that the indeterminate equation

$$(x + 1)^y - x^z = 1, \ x, \ y, \ z > 1 \tag{9.13}$$

has only one positive integer solution $x = 2$, $y = 2$, and $z = 3$.

Proof One First, equation (9.13) modulo $x + 1$ gets

$$-(-1)^z \equiv 1 \pmod{x+1},$$

thus z is odd. Factorize (9.13) into

$$(x + 1)^{y-1} = x^{z-1} - x^{z-2} + \cdots - x + 1,$$

which implies that x is even. Since if x is odd, then the right side of the above equation is the sum of z terms of odd numbers, which is odd. But the left side is even, a contradiction. Similarly, rewrite (9.13) as

$$(x+1)^{y-1} + (x+1)^{y-2} + \cdots + (x+1) + 1 = x^{z-1},$$

thus y is even, too.

Now let $x = 2x_1$ and $y = 2y_1$, then (9.13) can be factorized into

$$((x+1)^{y_1} - 1)((x+1)^{y_1} + 1) = x^z. \tag{9.14}$$

As x is even, the greatest common divisor of $(x+1)^{y_1} - 1$ and $(x+1)^{y_1} + 1$ is 2, and clearly $x \mid (x+1)^{y_1} - 1$. By this together with (9.14) we deduce

$$(x+1)^{y_1} - 1 = 2x_1^z, \quad (x+1)^{y_1} + 1 = 2^{z-1}.$$

Hence $2^{z-1} > 2x_1^z$, so $x_1 = 1$, that is, $x = 2$. Thus $y = 2$ and $z = 3$.

Proof Two We divide the proof into two steps. At first, we want to prove that x does not have odd prime divisors. We prove it by contradiction. Assume that there is an odd prime, such that $p \mid x$, says $x = p^a x_1$, where $a \geqslant 1$ and $p \nmid x_1$. By the binomial theorem, we can rewrite (9.13) as

$$xy + \sum_{i=2}^{y} \binom{y}{i} x^i = x^z. \tag{9.15}$$

Thus $x^2 \mid xy$, i.e., $x \mid y$. Hence $p \mid y$. Let $y = p^b y_1$, $p \nmid y_1$, then $b \geqslant a$. We will deduce a contradiction by comparing the powers of p on the two sides of (9.13).

For $2 \leqslant i \leqslant y$, let $p^c \parallel i$, then in

$$\binom{y}{i} x^i = \frac{y}{i} \binom{y-1}{i-1} x^i = \frac{p^b y_1}{i} \binom{y-1}{i-1} (p^a x_1)^i,$$

the power of p is at least $d = b + ai - c$. If $c = 0$, then $d > a + b$; if $c > 0$. Therefor from $p \geqslant 3$ we get $p^c > c + 1$. As $p^c \mid i$, we have $p^c \leqslant i$. Hence, $i > c + 1$, thus

$$d > b + a + c(a-1) \geqslant a + b.$$

Therefore we always have $d \geqslant a + b + 1$, thus $p^{a+b+1} \left| \binom{y}{i} x^i (2 \leqslant i \leqslant y) \right.$,

and

$$p^{a+b+1} \left| \sum_{i=2}^{y} \binom{y}{i} x^i \right. .$$

Also $p^{a+b} \parallel xy$, hence the power of p on the left side of (9.15) is $a + b$.

On the other hand, as $p^a \parallel x$, $p^{az} \parallel x^z$, i. e. , the power of p on the right side of (9.15) is az. But from the original equation (9.13) we get $z > y$, also $p^b \mid y$, so $y \geqslant p^b$, thus

$$az > ay \geqslant ap^b \geqslant a(b+1) \geqslant a+b.$$

Hence the powers of p on the two sides is not the same, this is impossible. Therefore in x there is no odd prime divisor. That is, x is a power of 2.

Assume $x = 2^k \ (k \geqslant 1)$. From the above proof we have $x \mid y$, thus y is even, say $y = 2y_1$. Equation (9.13) can be factorized into

$$((2^k+1)^{y_1} - 1)((2^k+1)^{y_1} + 1) = 2^{kz}.$$

Since the greatest common divisor of the two divisors on the left is 2, but the right side is a power of 2, thus we must have

$$(2^k+1)^{y_1} - 1 = 2; \ (2^k+1)^{y_1} + 1 = 2^{kz-1}.$$

Hence $k = y_1 = 1$, that is, $x = y = 2$, so $z = 3$.

Exercises

9. 1 Prove that the indeterminate equation

$$x^2 + 3xy - 2y^2 = 122$$

has no integer solution.

9. 2 Find all positive integers m and n, such that $\mid 12^m - 5^n \mid = 7$.

9. 3 Find all primes p, such that $2^p + 3^p$ is a k-th power of an integer (where $k \geqslant 2$).

9. 4 Prove that the indeterminate equation

$$5^x - 3^y = 2$$

has a unique positive integer solution $x = y = 1$.

Some problems of number theory in mathematical competition are difficult to handle. Solving them needs comprehensive and flexible use of knowledge and methods in the above chapters. In this chapter we introduce some of such problems.

Example 1   Assume that u is a positive integer. Prove that the equation

$$n! = u^x - u^y$$

has at most a finite number of positive integer solutions (n, x, y).

Proof   We can assume $u > 1$. The statement is equivalent to proving that the equation

$$n! = u^r (u^s - 1) \qquad (10.1)$$

has at most a finite number of positive integer solutions (n, r, s).

First, we note that for a given n, equation (10.1) has at most a finite number of solutions (r, s). In the following we prove that when n is sufficiently large, equation (10.1) has no solution, thus proving the above conclusion.

Fix a prime p with $p \nmid u$. We can assume that (10.1) has solution $n > p$ (otherwise nothing has to be proved), and $p^\alpha \parallel n!$. Then

$$\alpha = \sum_{l=1}^{\infty} \left[\frac{n}{p^l} \right] \geqslant \left[\frac{n}{p} \right] > an, \qquad (10.2)$$

where a is a (positive) constant depending only on p.

Assume that the order of u modulo p is d and $p^{k_0} \parallel (u^d - 1)$, then from Example 5 in Chapter 8, if $\alpha > k_0$ the order of u modulo p^α is $dp^{\alpha - k_0}$. As u and p are all fixed numbers, thus k_0 and d are all fixed numbers. If for sufficiently large n (10.1) has solutions, then by

(10.2) we have $\alpha > k_0$. But (10.1) implies

$$u^s \equiv 1 \pmod{p^\alpha},$$

thus by the properties of orders we get $dp^{\alpha-k_0} \mid s$. In particular, $s \geqslant dp^{\alpha-k_0}$. Hence,

$$u^s - 1 \geqslant u^{dp^{\alpha-k_0}} - 1 > u^{dp^{\alpha n-k_0}} - 1. \qquad (10.3)$$

But when n is sufficiently large, it is easy to see that the right side of (10.3) $\geqslant n^n - 1$. So (10.3) implies $u^s - 1 > n!$, and $u^r(u^s - 1) > n!$. Hence when n is sufficiently large (10.1) has no positive integer solutions (r, s). This completes the proof.

Remark 1   The proof of the second half is similar to the proof of Example 6 in Chapter 8. However, the problem there is more direct, and the problem here is more difficult. The key of this problem is calculating (10.1) by modulo p^α , and deducing that s is very large by using properties of order modulo p^α . Thus (10.1) does not hold when n is sufficiently large. The purpose we choose α such that $p^\alpha \parallel n!$ is to make α is very large (as n is very large) (cf. equation (10.2)).

Remark 2   If one has some knowledge of orders (involving infinite quantity), then after the result $s \geqslant dp^{\alpha-k_0}$ is deduced, one can easily find that (10.1) is not true. Put $b = u^{dp-k_0}$, then the inequality $b^{p^{\alpha n}} - 1 > n^n - 1$ discussed above becomes

$$p^{\alpha n} > n\log_b n.$$

When n is very large the left side is an exponential function of n (with base greater than 1), therefore larger than the right side which is a product of a linear function and a logarithmic function. (If setting $p^\alpha = 1 + \beta$ (where $\beta > 1$), and by using the binomial theorem one expands $(1 + \beta)^n$, then the above statement is clearly true.)

Example 2   Find all integers $n > 1$, such that $\dfrac{2^n + 1}{n^2}$ are integers.

Solution   It is easy to guess that $n = 3$ is a unique solution. Let us prove it.

We divide the proof into several steps. The first step is to consider the minimal prime divisor p of n, and from $n \mid (2^n + 1)$ we can deduce $p = 3$ (cf. Example 1 in Chapter 8). Hence we can assume

$$n = 3^m c, \ m \geqslant 1, \ 3 \nmid c. \tag{10.4}$$

In the second step, we prove $m = 1$. By $n^2 \mid (2^n + 1)$ we have $2^{3^m c} \equiv -1 \pmod{3^{2m}}$, thus

$$2^{2 \times 3^m c} \equiv 1 \pmod{3^{2m}}. \tag{10.5}$$

If $m \geqslant 2$, then according to Example 5 in Chapter 8 we get that the order of 2 modulo 3^{2m} is $2 \times 3^{2m-1}$. So (10.5) implies $2 \times 3^{2m-1} \mid 2 \times 3^m c$, i.e., $3^{m-1} \mid c$. Thus $3 \mid c$ (as $m \geqslant 2$), which contradicts $3 \nmid c$ in (10.4). So $m = 1$.

In the third step, we prove $c = 1$ in (10.4). This can be done similarly as the first step, i.e., Example 1 in Chapter 8:

If $c > 1$, let q be the least prime divisor of c, then

$$2^{3c} \equiv -1 \pmod{q}. \tag{10.6}$$

Let r be the order of 2 modulo q, by (10.6) we have $2^{6c} \equiv 1 \pmod{q}$. Further we also have $2^{q-1} \equiv 1 \pmod{q}$, so $r \mid 6c$ and $r \mid (q-1)$, thus $r \mid \gcd(6c, q-1)$. According to the choice of q we have $\gcd(q-1, c) = 1$, so $r \mid 6$. Also, $2^r \equiv 1 \pmod{q}$ implies $q = 3$ or 7. It is easy to see that $q = 3$ is impossible, and by (10.6), $q = 7$ is impossible, too. Thus $c = 1$, and $n = 3$.

Note that if we prove first $c = 1$ in (10.4), then it would not work well. Here the order of proof is very important. Besides, the identity $m = 1$ in the second step can also be proved by comparing the powers of primes as follows.

It follows from the binomial theorem that

$$2^n + 1 = (3-1)^n + 1 = 3n + \sum_{k=2}^{n} (-1)^k \binom{n}{k} 3^k. \tag{10.7}$$

Let $3^\alpha \parallel k!$. Then

$$\alpha = \sum_{l=1}^{\infty} \left[\frac{k}{3^l}\right] < \sum_{l=1}^{\infty} \frac{k}{3^l} = \frac{k}{2}.$$

Thus by

$$\binom{n}{k}3^k = \frac{n(n-1)\cdots(n-k+1)}{k!}3^k$$

we have that $\binom{n}{k}3^k$ is divisible by 3^β, and β satisfies (note that $k \geq 2$)

$$\beta \geq k + m - \alpha > k + m - \frac{k}{2} \geq m + 1,$$

so $\beta \geq m + 2$. Thus the sum on the right side of (10.7) is divisible by 3^{m+2}. If $m > 1$, then $2m \geq m + 2$, so it follows from $3^{2m} \mid (2^n + 1)$ and (10.7) that $3^{m+2} \mid 3n$, i.e., $3^{m+1} \mid n$, in contradiction with (10.4). Thus $m = 1$.

In the following Example 3 we can also apply the method of comparing the powers of prime.

Example 3   Prove that for each $n > 1$, the equation

$$\frac{x^n}{n!} + \frac{x^{n-1}}{(n-1)!} + \cdots + \frac{x^2}{2!} + \frac{x}{1!} + 1 = 0$$

has no rational roots.

Proof   Assume that a is a rational root of the equation. Then

$$a^n + \frac{n!}{(n-1)!}a^{n-1} + \cdots + \frac{n!}{k!}a^k + \cdots + \frac{n!}{1!}a + n! = 0. \quad (10.8)$$

Thus a is a rational root of a polynomial with integer coefficients and the leading coefficient 1. So a is an integer (cf. Exercise 2.4).

Since $n > 1$, n has a prime divisor (this fundamental fact has been used many times). As $n \left| \dfrac{n!}{k!} \right.$ ($k = 0, 1, \ldots, n-1$), by (10.8) we have $p \mid a^n$, thus the prime p divides a. Now compare the powers containing p in each term on the left side of (10.8). Since the number of p occurring in $k!$ is

$$\sum_{l=1}^{\infty}\left[\frac{k}{p^l}\right] < \sum_{l=1}^{\infty}\frac{k}{p^l} < k,$$

$p^k \nmid k!$ ($k \geq 1$). Assume that $p^r \parallel n!$, then it follows from $p^k \mid a^k$ and

$p^k \nmid k!$ that $p^{r+1} \left| \dfrac{n!}{k!} a^k \right.$ $(k = 1, 2, \ldots, n)$. Thus (10.8) implies that $p^{r+1} \mid n!$, in contradiction with the definition of r.

Remark By a more complicated method we can prove that the polynomial with rational coefficients

$$\frac{x^n}{n!} + \frac{x^{n-1}}{(n-1)!} + \cdots + \frac{x^2}{2!} + \frac{x}{1!} + 1$$

is irreducible over the rational number field, that is, it cannot be decomposed into a product of two non-constant polynomials with rational coefficients. Example 3 is a special case of this conclusion: the polynomial in question has no rational divisor of degree one.

In mathematical competition, some problems of number theory involving polynomials occur frequently. Let us show some examples.

Example 4 Assume that $n > 1$, $x_1, \ldots, x_n$ are n real numbers, and their product is denoted by A. If $A - x_i$ are all odd for $i = 1, \ldots, n$, prove that each x_i is irrational.

Proof We prove it by contradiction. If there is i such that x_i is a rational number, then as $A - x_i$ is odd, A must be a rational number. Write $A - x_i = a_i (i = 1, \ldots, n)$. Then by $x_1 \cdots x_n = A$ we have

$$(A - a_1) \cdots (A - a_n) = A. \tag{10.9}$$

Since a_i are all (odd) integers, A satisfies an equation with integer coefficients and the leading coefficient one. So the rational number A is an integer. But on the other hand, whether A is odd or even, the parities of the two sides of (10.9) are different. Thus (10.9) does not hold, a contradiction! Hence, every x_i is an irrational number.

Rational roots of a polynomial with integer coefficients and the leading coefficient ± 1 must be integers. This simple result has many uses. Sometimes, we can use it to prove that a rational number is actually an integer.

Example 5 Let a, b and c be integers, and $f(x) = x^3 + ax^2 + bx + c$. Prove that there are infinitely many positive integers n, such that $f(n)$ is not a perfect square.

Proof　We prove that for any positive integer $n \equiv 1 \pmod 4$, among the four numbers $f(n)$, $f(n+1)$, $f(n+2)$, $f(n+3)$ there is at least one number which is not a perfect square, which in turn will prove the conclusion in the problem.

It is easy to get

$$f(n) \equiv 1 + a + b + c \pmod 4,$$
$$f(n+1) \equiv 2b + c \pmod 4,$$
$$f(n+2) \equiv -1 + a - b + c \pmod 4,$$
$$f(n+3) \equiv c \pmod 4.$$

Eliminating a and c, we have

$$f(n+1) - f(n+3) \equiv 2b, \ f(n) - f(n+2) \equiv 2b + 2 \pmod 4.$$

Hence, either $f(n+1) - f(n+3) \equiv 2 \pmod 4$, or $f(n) - f(n+2) \equiv 2 \pmod 4$. Since perfect squares modulo 4 are congruent to 0 or 1, either one of $f(n+1)$ and $f(n+3)$ is not a perfect square, or one of $f(n)$ and $f(n+2)$ is not a perfect square. Thus among $f(n)$, $f(n+1)$, $f(n+2)$, $f(n+3)$ there is at least one which is not a perfect square.

Example 6　Assume that $p(x)$ is a polynomial with integer coefficients, and for any $n \geqslant 1$, $p(n) > n$. Define $x_1 = 1$, $x_2 = p(x_1)$, $\ldots$, $x_n = p(x_{n-1})$ $(n \geqslant 2)$. If for any positive integer N, in the sequence $\{x_n\}$ $(n \geqslant 1)$ there is a term which is divisible by N, prove that $p(x) = x + 1$.

Proof　We prove it in two steps. At first, we will prove that for any fixed $m > 1$, the sequence $\{x_n\}$ modulo $x_m - 1$ is a periodic sequence. Obviously, $x_m \equiv 1 = x_1 \pmod{x_m - 1}$. As $p(x)$ is a polynomial with integer coefficients, for any integers u and v $(u \neq v)$, we have $(u - v) \mid (p(u) - p(v))$, that is,

$$p(u) \equiv p(v) \pmod{u - v}.$$

In the above equation take $u = x_m$, $v = x_1 = 1$, we get $x_{m+1} \equiv x_2 \pmod{x_m - 1}$. Similarly, $x_{m+2} \equiv x_3$, $x_{m+3} \equiv x_4$, $\ldots$ $\pmod{x_m - 1}$, thus $\{x_n\}$ modulo $x_m - 1$ is a periodic sequence $x_1, \ldots, x_{m-1}, x_1, \ldots, x_{m-1}, \ldots$.

In the second step, we will prove that

$$x_m - 1 = x_{m-1}. \tag{10.10}$$

From the hypothesis it follows that for $N = x_m - 1$ there is x_k such that $(x_m - 1) \mid x_k$. From the above results we can assume $1 \leqslant k \leqslant m - 1$. Moreover, $p(x_{m-1}) > x_{m-1}$, so $x_m - 1 \geqslant x_{m-1}$. Hence k must be $m - 1$, that is, $(x_m - 1) \mid x_{m-1}$. Thus, $x_{m-1} \geqslant x_m - 1$, which implies that (10.10) holds.

In view of (10.10), $p(x_{m-1}) - 1 = x_{m-1}$, and m is any integer greater than 1, which means that $p(x) = x + 1$ has infinitely many different roots. So $p(x)$ must be the polynomial $x + 1$. This completes the proof.

From an arithmetic property of a polynomial (with integer coefficients) we deduce its algebraic properties, it is an interesting topic of number theory. Example 6 is such an example, so is the following Example 7.

Example 7　Suppose that $f(x)$ is a quadratic real polynomial. If for any positive integer n, $f(n)$ is a square of an integer. Prove that $f(x)$ is a square of a one-degree polynomial with integer coefficients.

Proof　This problem is not easy, but there are several completely different solutions. Here we give a comparatively easy solution based on limited knowledge of sequences.

Set $f(x) = ax^2 + bx + c$, $a_n = f(n)$ $(n \geqslant 1)$, then it is easy to prove that

$$
\begin{aligned}
\sqrt{a_n} - \sqrt{a_{n-1}} &= \frac{a_n - a_{n-1}}{\sqrt{a_n} + \sqrt{a_{n-1}}} \\
&= \frac{2an - a + b}{\sqrt{an^2 + bn + c} + \sqrt{an^2 + (-2a + b)n + a - b + c}} \\
&= \frac{2a + \dfrac{b - a}{n}}{\sqrt{a + \dfrac{b}{n} + \dfrac{c}{n^2}} + \sqrt{a + \dfrac{b - 2a}{n} + \dfrac{a - b + c}{n^2}}}.
\end{aligned}
$$

Hence, when $n \to \infty$, $\sqrt{a_n} - \sqrt{a_{n-1}}$ has a limit, and the limit is

$\dfrac{2a}{\sqrt{a}+\sqrt{a}} = \sqrt{a}$. But we know that $\sqrt{a_n}$ are all integers, $\{\sqrt{a_n} - \sqrt{a_{n-1}}\}$ $(n \geqslant 2)$ is a sequence of integers, hence its limit value $\sqrt{a}$ is an integer, and when n is sufficiently large all terms $\sqrt{a_n} - \sqrt{a_{n-1}}$ equal the limit $\sqrt{a}$, that is, there is a fixed positive integer k, such that

$$\sqrt{a_n} - \sqrt{a_{n-1}} = \sqrt{a}, \text{ for all } n \geqslant k+1.$$

Now assume that m is any integer greater than k, summing the above equations for $n = k+1, \ldots, m$, we get $\sqrt{a_m} = \sqrt{a_k} + (m-k)\sqrt{a}$, that is,

$$a_m = (m\sqrt{a} + \sqrt{a_k} - k\sqrt{a})^2. \tag{10.11}$$

Write $\alpha = \sqrt{a}$, and $\beta = \sqrt{a_k} - k\sqrt{a}$, then α, β are fixed integers and independent of m. Thus from (10.11) it follows that all integers m greater than k are roots of the polynomial

$$f(x) - (\alpha x + \beta)^2.$$

Hence, this polynomial must be zero, that is, $f(x) = (\alpha x + \beta)^2$.

In mathematical competition we often come across some problems in number theory with combinatorial flavour. For these problems we do not need much pre-reguisite, but they are very versatile. We have met them in former chapters. Now we show more.

Example 8　Assume that $n > 1$ and the sum of n positive integers is $2n$. Prove that among these integers we can choose some numbers, such that their sum is n, except for the given numbers satisfying one of the following conditions:

(1) There is a number which is $n+1$, and the others are 1.

(2) When n is odd, all numbers equal 2.

Proof　Assume that the given numbers are $0 < a_1 \leqslant a_2 \leqslant \cdots \leqslant a_n$. Set $S_k = a_1 + \cdots + a_k (k = 1, \ldots, n-1)$. Then among the following numbers

$$0, a_1 - a_n, S_1, \ldots, S_{n-1}$$

there must be two numbers which are the same after modulo n. We discuss this by dividing it into four cases:

(i) Assume that there is $S_k (1 \leqslant k \leqslant n-1)$, such that $S_k \equiv 0 \pmod{n}$. In this case

$$1 \leqslant S_k \leqslant a_1 + \cdots + a_n - a_{k+1} \leqslant 2n-1, \qquad (10.12)$$

so $S_k = n$.

(ii) Suppose that there are S_i and $S_j (1 \leqslant i < j \leqslant n-1)$, satisfying $S_i \equiv S_j \pmod{n}$, then from (10.12) we have $1 \leqslant S_j - S_i \leqslant 2n-1$, so $S_j - S_i = n$, that is,

$$a_{i+1} + \cdots + a_j = n.$$

(iii) Suppose that there is $S_k (1 \leqslant k \leqslant n-1)$, such that $S_k \equiv a_1 - a_n \pmod{n}$. If $k = 1$, then $a_n \equiv 0 \pmod{n}$. But $a_1, \ldots, a_{n-1}$ are all positive integers, so $a_1 + \cdots + a_{n-1} \geqslant n-1$. Thus

$$a_n = 2n - (a_1 + \cdots + a_{n-1}) \leqslant n+1. \qquad (10.13)$$

Hence $a_n = n$, in this case the result holds. If $k > 1$, then

$$a_2 + \cdots + a_k + a_n \equiv 0 \pmod{n}.$$

However, the left side of the above equation is a positive integer less than $a_1 + \cdots + a_n = 2n$, so

$$a_2 + \cdots + a_k + a_n = n.$$

(iv) Assume that $a_1 - a_n \equiv 0 \pmod{n}$. We have proved that $a_n \leqslant n + 1$ (see Equation (10.13)). The sum of $n - 1$ positive integers $a_1, \ldots, a_{n-1}$ is equal to $2n - a_n = n - 1$ provided $a_n = n + 1$. Thus they are all equal to 1. This is the case (1) ruled out in the problem.

Assume that $a_n \leqslant n$, then $0 \leqslant a_n - a_1 \leqslant n - 1$, together with $a_n - a_1 \equiv 0 \pmod{n}$ we deduce $a_n = \cdots = a_2 = a_1 = 2$. When n is odd this is the case (2) ruled out in the problem. If n is even, then the sum of any $\dfrac{n}{2}$ numbers a_i is equal to n.

The proof of Example 8 can be summarized as follows. Since the sum of all given numbers is $2n$, we only need to prove that the sum of some (not all) numbers is a multiple of n, then this sum is just n, and the latter is a problem we can deal with using congruence.

Example 9   Let p be a prime. For given $p + 1$ different positive integers, prove that we can choose one pair of numbers among them, such that after the larger number is divided by the greatest common divisor of the two numbers, the integer quotient is not less than $p + 1$.

Proof   After the given $p + 1$ numbers are divided by the greatest common divisor of them, obviously this does not affect the conclusion of the problem. Thus we can assume that these $p + 1$ numbers are relatively prime. In particular, among them there is a number which is not divisible by p. Denote these $p + 1$ numbers by

$$x_1, \ldots, x_k, \ x_{k+1} = p^{l_{k+1}} y_{k+1}, \ldots, \ x_{p+1} = p^{l_{p+1}} y_{p+1},$$

where $x_1, \ldots, x_k$ are distinct and relatively prime to p $(k \geqslant 1)$, $l_{k+1}, \ldots,$ l_{p+1} are positive integers, $y_{k+1}, \ldots, y_{p+1}$ are positive integers and not divisible by p.

Among $p + 1$ numbers

$$x_1, \ldots, x_k, y_{k+1}, \ldots, y_{p+1}, \tag{10.14}$$

there must be two numbers which are congruent modulo p. We discuss this by dividing it into three cases.

(1) Among (10.14) there are at least three numbers which are the same. In this case the statement can be proved easily. If $y_r = y_s = y_t$, then $p^{l_r}, p^{l_s}, p^{l_t}$ are distinct, where the greatest number is at least p^2 times of the least numbers, we can assume that $p^{l_r} \geqslant p^2 \cdot p^{l_t}$, then x_r and x_t meet the requirement. If $y_r = y_s = x_t (1 \leqslant t \leqslant k)$, we can assume that $l_r > l_s$, then $l_r \geqslant 2$, thus x_r and x_t meet the requirement.

(2) Among (10.14) there are two pairs of numbers, such that the two numbers in each pair are the same. If $y_i = y_j$, we have $y_r = y_s$. Then when $\mid l_i - l_j \mid \geqslant 2$ or $\mid l_r - l_s \mid \geqslant 2$, by a similar discussion we get the result. When $\mid l_i - l_j \mid \leqslant 1$ and $\mid l_r - l_s \mid \leqslant 1$, we can denote $x_i, x_j,$ x_r, x_s by a, ap, b, bp, respectively, and $a < b$. In this case,

$$\frac{bp}{\gcd(a, bp)} \geqslant \frac{bp}{a} > p,$$

so the integer $\dfrac{bp}{\gcd(a,\,bp)} \geqslant p + 1$.

If $x_i = y_r$, $x_j = y_s (1 \leqslant i,\, j \leqslant k)$, by the same argument we can also get the result.

(3) Among (10. 14) there are just two numbers which are the same, this is the case only when $y_r = y_s$ or $x_i = y_r (1 \leqslant i \leqslant k)$. In this case we can delete y_r in (10.14), then the rest of the p numbers are distinct, but there are still two numbers which are the same modulo p. There are still three other possibilities:

(i) Let $y_r \equiv y_s \pmod{p}$. We can assume that $y_r > y_s$. If $l_r > l_s$, the result is obviously true. If $l_r \leqslant l_s$, write $y_r = y_s + n$, then $n > 0$, and $p \mid n$. Let $\gcd(y_r,\, y_s) = d$, then $p \nmid d$, thus $\gcd(x_r,\, x_s) = p^{l_r} d$, we have (note that $d \mid n$, $p \mid n$, and $p \nmid d$)

$$\frac{x_r}{\gcd(x_r,\, x_s)} = \frac{y_r}{d} = \frac{y_s}{d} + \frac{n}{d} \geqslant 1 + p.$$

Therefore, the larger one between x_r and x_s divides their greatest common divisor, the quotient is at least $p + 1$.

(ii) Let $x_r \equiv x_s \pmod{p}$ $(1 \leqslant r < s \leqslant k)$. This can be solved similarly to (i).

(iii) Let $x_r \equiv y_s \pmod{p}$ $(1 \leqslant r \leqslant k)$. If $y_s > x_r$, then the result is clearly true. If $y_s < x_r$, let $x_r = y_s + n$, then $n > 0$, and $p \mid n$. Let $\gcd(x_r,\, y_s) = d$, then $p \nmid d$. Thus $\gcd(x_r,\, x_s) = \gcd(x_r,\, p^{l_s} y_s) = d$, therefore

$$\frac{x_r}{\gcd(x_r,\, x_s)} = \frac{y_s}{d} + \frac{n}{d} \geqslant 1 + p.$$

Hence, after the larger number between x_r and x_s is divided by their greatest common divisor, the quotient is not less than $p + 1$.

This completes the proof.

We note that if the $p + 1$ integers in Example 9 are replaced by p integers, then the conclusion does not hold. For instance, clearly, among the p numbers 1, 2, ..., p there do not exist two numbers which meet the requirement.

Example 10   Let S be a subset of $\{1, 2, \ldots, 2^m n\}$, and $|S|$ the number of elements in S, such that $|S| \geqslant (2^m - 1)n + 1$. Prove that in S there are $m + 1$ distinct numbers $a_0, \ldots, a_m$, such that $a_{i-1} \mid a_i$ ($i = 1, \ldots, m$).

Proof   Each positive integer a can be expressed in the form $2^u k$, where $u \geqslant 0$, and k is odd. We call k the odd part of a, and if the odd part of a is not larger than n, then n is called a *good number*. Our proof is based on the estimate of lower bound of the number of good numbers in S.

Suppose that in the interval $[1, n]$ there are t odd numbers (in fact, t equals $\left[\dfrac{n+1}{2}\right]$, but we do not need this result). Let k be one such number, then the number of integers u satisfying $n < 2^u k \leqslant 2^m n$ is just m. For this we note that if an integer v satisfying $2^{v-1} \leqslant \dfrac{n}{k} < 2^v$, then $2^v k, 2^{v+1} k, \ldots, 2^{v-1+m} k$ are all required numbers, that is, in the interval $(n, 2^m n]$ there are m numbers such that their odd part is k, so among them there are only mt good numbers. Hence in this interval there are $2^m n - n - mt$ non-good numbers. Thus in S the number of good numbers $\geqslant |S| - (2^m n - n - mt) = mt + 1$.

Assume that $k_1, \ldots, k_t$ are all odd numbers in $[1, n]$, and there are only x_i numbers whose odd part are k_i ($k = 1, \ldots, t$), then from the result in the last section it follows that the number of good numbers in S is

$$x_1 + \cdots + x_t \geqslant mt + 1.$$

Thus there is x_i ($1 \leqslant i \leqslant t$), such that $x_i \geqslant m + 1$, that is, there are at least $m + 1$ integers in S whose odd parts are all k_i, these numbers are (from small to large) $a_0, a_1, \ldots, a_m$, which are the $m + 1$ required numbers. This completes the proof.

Remark 1   When $m = 1$, we get a well-known result, the proof here is a generalization of this (ordinary) result. There are other solutions of this problem. For instance, by induction on m or n.

Interested readers can try it themselves.

Remark 2   The set $S = \{n + 1, \ldots, 2^m n\}$ shows that if S in Example 10 satisfies $\mid S \mid = (2^m - 1)n$, then the conclusion is not always true. This is so because if $a_0, \ldots, a_m$ meet the requirement, then $a_m \geqslant 2^m a_0$, hence $a_m \geqslant 2^m (n + 1)$, which is impossible.

Example 11   Let A be a set with n positive integers $(n \geqslant 2)$. Prove that there is a subset B of A satisfying $\mid B \mid > \dfrac{n}{3}$, and $x + y \notin B$ for any $x, y \in B$.

Proof   We denote all numbers in A by $a_1, \ldots, a_n$. From Exercise 3. 2 we know that there are infinitely many primes which modulo 3 are -1. Choose one such prime $p > a_i (1 \leqslant i \leqslant n)$. Assume that $p = 3k - 1$. Let us consider the following pn numbers (with p rows and n columns)

$$
\begin{aligned}
&a_1, a_2, \ldots, a_n; \\
&2a_1, 2a_2, \ldots, 2a_n; \\
&\cdots\cdots\cdots\cdots \\
&pa_1, pa_2, \ldots, pa_n.
\end{aligned}
\qquad (10.15)
$$

As $p > a_i$, $\gcd(p, a_i) = 1$. Hence every column in (10.15) is a complete system modulo p (see (10) in Chapter 6), thus for every j $(0 \leqslant j < p)$, there are in total n numbers in (10.15) which modulo p are j. Therefore, there are kn numbers which modulo p are one of $k, k + 1, \ldots, 2k - 1$.

Assume that in the i-th row in (10.15) there are x_i numbers which modulo p are one of $k, k + 1, \ldots, 2k - 1$. Then according to the above argument we have

$$
x_1 + \cdots + x_p = kn.
$$

So there is x_i, such that

$$
x_i \geqslant \frac{kn}{p} = \frac{kn}{3k - 1} > \frac{n}{3},
$$

i. e., there is l $(1 \leqslant l \leqslant p)$, such that the number of elements in la_1,

$la_2, \ldots, la_n$ which modulo p are one of k, $k+1$, $\ldots$, $2k-1$ is larger than $\frac{n}{3}$. We put

$$B = \{a \in A \mid la \text{ modulo } p \text{ is one of } k, k+1, \ldots, 2k-1\},$$

then B is a required set, because for any x, $y \in B$, it is easy to obtain that the remainder of $l (x+y)(= lx+ly)$ modulo p is either $\geq 2k$ or $\leq 2k-1$, then $x+y \notin B$.

This solution is provided by N. Alon, a famous Israel mathematician. It is very clever and you may enjoy it.

At the end of this chapter, we show two problems which can be solved by construction method.

Example 12   Let $n \geq 2$. Prove that there exist n distinct positive integers possessing the following properties.

(1) These numbers are pairwise relatively prime;

(2) The sums of any k $(2 \leq k \leq n)$ numbers among them are composite.

Proof   If $n = 2$ the statement is obviously true. Suppose that there have been n positive integers which meet the requirements. According to this fact we will construct $n+1$ required numbers.

As there are infinitely many primes, we can choose 2^n-1 distinct primes $p_i (1 \leq i \leq 2^n-1)$ which are all relatively prime with $a_1 a_2 \cdots a_n$. Denote by $S_j (1 \leq j \leq 2^n-1)$ the 2^n-1 sums of k numbers $(1 \leq k \leq n)$ chosen from $a_1, a_2, \ldots, a_n$, where the sums are numbers $a_i (1 \leq i \leq n)$ when $k = 1$.

Since $\gcd(p_i, a_1 \cdots a_n) = 1$, there exists b_i such that $a_1 \cdots a_n \cdot b_i \equiv 1 \pmod{p_i}$ $(1 \leq i \leq 2^n-1)$. By the Chinese remainder theorem, the system of congruences

$$x \equiv -b_i - b_i S_i \pmod{p_i}, \ 1 \leq i \leq 2^n-1 \qquad (10.16)$$

has infinitely many solutions x of positive integers. Choose one solution $x_0 > p_i (1 \leq i \leq 2^n-1)$, multiply $a_1 \cdots a_n$ on the two sides of (10.16), we obtain

$$a_1 \cdots a_n x_0 + 1 + S_i \equiv 0 \pmod{p_i}, \quad 1 \leqslant i \leqslant 2^n - 1. \quad (10.17)$$

Set $a_{n+1} = a_1 \cdots a_n x_0 + 1$, then the $n+1$ numbers $a_1, \ldots, a_n, a_{n+1}$ meet the requirements. This is because $x_0 > p_i$, $a_{n+1} + S_i > p_i$. But (10.17) implies that $a_{n+1} + S_i$ has divisor p_i, so for any i, $a_{n+1} + S_i$ is a composite number. By the construction of a_{n+1}, it is relatively prime with every $a_i (1 \leqslant i \leqslant n)$, completing the inductive construction.

The key of the above solution is: if $a_1, \ldots, a_n$ are given, we want to choose one value of a parameter x, such that the number $a_1 \cdots a_n x + 1$ can be constructed to be a_{n+1}. The main benefit constructing such kind of numbers is that the conditions $\gcd(a_{n+1}, a_i) = 1$ $(1 \leqslant i \leqslant n)$ hold automatically.

Elements that met the requirements are usually not unique, we can try to choose some elements with some special properties, that is the elements satisfying some suitable sufficient conditions to meet part of requirements in question. This kind of methods has many applications in constructive proofs.

We can solve this problem by using the following constructive method (more directly): choose $a_i = i \cdot n! + 1$, then $a_1, \ldots, a_n$ meet the requirements. This is because:

At first, for $i \neq j$, we have $\gcd(a_i, a_j) = 1$. This is because if $\gcd(a_i, a_j) = d$, then $j a_i - i a_j$ is a multiple of d, that is, $d \mid (i - j)$. But $1 \leqslant |i - j| < n$, so $d \mid n!$, thus from $d \mid a_i$ it follows $d = 1$.

Moreover, the sum of any k $(2 \leqslant k \leqslant n)$ numbers a_i has a form $m \cdot n! + k$ (m is some integer), this number has obviously a proper divisor k, thus it is not prime.

Example 13 Find all positive integers, such that there exists a positive integer n satisfying

$$\frac{\tau(n^2)}{\tau(n)} = k, \quad (10.18)$$

where $\tau(n)$ is the number of positive divisors of n.

Solution By the formula of $\tau(n)$ in (6), Chapter 3, $\tau(n^2)$ is odd, therefore k satisfying (10.18) is odd. We will prove that every positive odd number k meets the requirement.

Clearly, $k = 1$ meets the requirement. For $k > 1$, by the formula of $\tau(n)$ we obtain that the problem is equivalent to proving that there exist positive integers $\alpha, \beta, \ldots, \gamma$, such that

$$\frac{(2\alpha + 1)}{\alpha + 1} \cdot \frac{(2\beta + 1)}{\beta + 1} \cdot \cdots \cdot \frac{(2\gamma + 1)}{\gamma + 1} = k. \qquad (10.19)$$

Now suppose that all odd numbers less than k meet the requirement. For the odd number k, we can assume $k = 2^l m - 1$, where $l \geqslant 1$, and m is an odd number. Since $k > 1$, we have $m < k$, so by the assumption in induction, there exist $\alpha', \beta', \ldots, \gamma'$, such that

$$\frac{(2\alpha' + 1)}{\alpha' + 1} \cdot \frac{(2\beta' + 1)}{\beta' + 1} \cdot \cdots \cdot \frac{(2\gamma' + 1)}{\gamma' + 1} = m. \qquad (10.20)$$

Now we choose two integers to be determined $x \geqslant 1$ and $u \geqslant 0$, satisfying $2^u \mid x$, and

$$\frac{2x + 1}{x + 1} \cdot \frac{2 \cdot \dfrac{x}{2} + 1}{\dfrac{x}{2} + 1} \cdot \cdots \cdot \frac{2 \cdot \dfrac{x}{2^n} + 1}{\dfrac{x}{2^n} + 1} = \frac{k}{m}. \qquad (10.21)$$

Obviously, if we can find out u and v satisfying the above requirement, then multipling (10.20) and (10.21), we can get an expression of k in a form such as (10.19). Hence we have proved that k meets the requirement, which completes the construction inductively.

In fact, (10.21) can be rewritten as

$$\frac{2x + 1}{\dfrac{x}{2^n} + 1} = \frac{k}{m},$$

which implies that (note that $k = 2^l m - 1$)

$$x = \frac{2^u(k - m)}{2^{u+1} m - 2^l m + 1}.$$

Hence, if we choose $u = l - 1$, then $u \geqslant 0$, the corresponding $x =$

$2^{l-1}(k - m)$ is a positive integer, and is divisible by $2^{l-1}(= 2^u)$, completing the proof.

Exercises

10. 1 Prove that for any integer $a \geqslant 3$, there exist infinitely many positive integers n, such that $a^n - 1$ is divisible by n.

10. 2 Suppose that $n_1, \ldots, n_k$ are positive integers, satisfying the following properties

$$n_1 \mid (2^{n_2} - 1), n_2 \mid (2^{n_3} - 1), \ldots, n_k \mid (2^{n_1} - 1).$$

Prove that $n_1 = \cdots = n_k = 1$.

10. 3 Suppose that integers a and b satisfy $a^2 b \mid (a^3 + b^3)$. Prove that $a = b$.

10. 4 Prove that the indeterminate equation $x^n + 1 = y^{n+1}$ has no positive integer solution (x, y, n), where $\gcd(x, n+1) = 1$, $n > 1$.

10. 5 Let a and b be positive rational numbers and $a \neq b$. If there exist infinitely many positive integers n, such that $a^n - b^n$ are integers, then a and b are integers.

10. 6 Assume that $n \geqslant 4$ is an integer, and $a_1, \ldots, a_n$ are distinct positive integers less than $2n$. Prove that we can choose certain members from these numbers, such that their sum is divisible by $2n$.

Appendix: Answers of Exercises

1.1 Among $1, 2, \ldots, n$, the numbers $k, 2k, \ldots, dk$ are divisible by k, where the positive integer d satisfies $dk \leqslant n$ and $(d + 1)k > n$. Hence, $\frac{n}{k} - 1 < d \leqslant \frac{n}{k}$, that is, $d = \left[\frac{n}{k}\right]$. So there are $\left[\frac{n}{k}\right]$ numbers which are divisible by k among the numbers considered.

1.2 Since numbers of mushrooms picked up by each child are the same, the number of children $n + 11$ divides the number of mushrooms

$$n^2 + 9n - 2 = (n + 11)(n - 2) + 20.$$

Hence, $n + 11$ divides 20. As $n + 11 > 11$, n must be 9. Therefore, there are more girls than boys.

1.3 We know that

$$n - T(n) = (a_0 - a_0) + (10a_1 + a_1) + \cdots + (a_k \times 10^k - (-1)^k a_k).$$

It is easy to see that for $i = 0, 1, \ldots, k$, $a_i \times 10^i - (-1)^i a_i$ is divisible by 11 (depending on whether i is even or odd, we apply factorization (5) or (6), respectively). Hence $n - T(n)$ is divisible by 11. Thus the divisibility condition for 11 is necessary and sufficient.

1.4 Let $a_1, \ldots, a_n$ be integers with the given properties, and A their product. n divides $\frac{A}{a_i} - a_i$ for $1 \leqslant i \leqslant n$, hence divides

$$a_i\left(\frac{A}{a_i} - a_i\right) = A - a_i^2.$$

So n divides the sum of these numbers

$$(A - a_1^2) + \cdots + (A - a_n^2) = nA - (a_1^2 + \cdots + a_n^2).$$

Thus n divides $a_1^2 + \cdots + a_n^2$.

1.5 If a, b, c are all divisible by $ad - bc$, then $(ad - bc)^2$ divides ad and bc, and so divides $ad - bc$, which implies $|ad - bc| = 1$, contradicting the given condition $ad - bc > 1$.

2.1 We have $4(9n + 4) - 3(12n + 5) = 1$.

2.2 Let $d = \gcd(2^m - 1, 2^n + 1)$. Then $2^m - 1 = du$, $2^n + 1 = dv$, where u and v are integers. It is easy to get $(du + 1)^n = (dv - 1)^m$, by expanding the two sides (note that m is odd), we get $dA + 1 = dB - 1$ (A and B are integers), hence $d \mid 2$, that is, $d = 1$ or 2. But obviously, d must be 1.

2.3 Since $\gcd(a, b) = 1$, we have $\gcd(a^2, b) = 1$. Hence $\gcd(a^2 + b^2, b) = 1$. Similarly, $\gcd(a^2 + b^2, a) = 1$. Thus $\gcd(a^2 + b^2, ab) = 1$ (by using (6) of this chapter).

2.4 Let rational number $\dfrac{p}{q}$ (with $\gcd(p, q) = 1$) be a root of polynomial

$$f(x) = x^n + a_1 x^{n-1} + \cdots + a_{n-1} x + a_n$$

with integer coefficients. From $f\left(\dfrac{p}{q}\right) = 0$ it follows that

$$p^n + a_1 p^{n-1} q + \cdots + a_{n-1} pq^{n-1} + a_n q^n = 0.$$

Since $a_1 p^{n-1} q$, $\ldots$, $a_{n-1} pq^{n-1}$, $a_n q^n$ are all divisible by q, we have $q \mid p^n$. But $\gcd(p, q) = 1$, hence $\gcd(q, p^n) = 1$. Thus $q = \pm 1$, that is, the rational number $\dfrac{p}{q}$ is an integer.

2.5 According to (10) of Chapter 2, the given condition is

$$\frac{(m + k)m}{\gcd(m + k, m)} = \frac{(n + k)n}{\gcd(n + k, n)}.$$

Since

$$\gcd(m + k, m) = \gcd(m, k),$$
$$\gcd(n + k, n) = \gcd(n, k),$$

from the above equation we have

$$\frac{(m+k)m}{\gcd(m,k)} = \frac{(n+k)n}{\gcd(n,k)}. \tag{A.1}$$

Suppose that $\gcd(m, k) = d_1$. Then $m = m_1 d_1$, $k = k_1 d_1$, where $\gcd(m_1, k_1) = 1$. Next, suppose that $\gcd(n, k) = d_2$. Then $n = n_1 d_2$, $k = k_2 d_2$, where $\gcd(n_1, k_2) = 1$. Thus equation (A.1) becomes

$$(m_1 + k_1)m_1 d_1 = (n_1 + k_2)n_1 d_2.$$

Multiplying two sides of the above equation by k_1, and using $k_1 d_1 = k_2 d_2 (= k)$, we get

$$(m_1 + k_1)m_1 k_2 = (n_1 + k_2)n_1 k_1.$$

The left side of the above equation is a multiple of k_2. Thus k_2 divides the right side, that is, $k_2 \mid k_1 n_1^2$. But $\gcd(k_2, n_1) = 1$, so $\gcd(k_2, n_1^2) = 1$, thus $k_2 \mid k_1$. Similarly, $k_1 \mid k_2$. Therefore, $k_1 = k_2$, i.e., $\gcd(m, k) = \gcd(n, k)$. Due to (A.1), $(m+k)m = (n+k)n$, thus we get $m = n$.

3.1   It is easy to verify that $(n+1)! + 2$, $(n+1)! + 3$, ..., $(n+1)! + (n+1)$ are n consecutive composite numbers.

3.2   We can prove it by a method similar to the method by which Euclid proved that there are infinitely many primes. Assume that there are only finitely many primes of the form $4k - 1$, we set all of them to be $p_1, \ldots, p_m$. Consider the number $N = 4p_1 \cdots p_m - 1$. Obviously, $N > 1$, and N has prime divisors. Moreover, a product of two primes of the form $4k + 1$ is also a number of the form $4k + 1$. But N is of the form $4k - 1$, so N must have prime divisor p of the form $4k - 1$. By the above assumption, p is one of $p_1, \ldots, p_m$. Hence $N - 4p_1 \cdots p_m$ is divisible by p, namely $p \mid 1$, a contradiction. Similarly we can prove that there are infinitely many primes of the form $6k - 1$.

3.3   Put $m = 9k^3 (k = 1, 3, \ldots)$, then

$$8^m + 9m^2 = (2^m)^3 + (9k^2)^3.$$

It is easy to verify that it has a proper divisor $2^m + 9k^2$.

3.4   Prove by contradiction. Suppose that there is a required set of numbers a, b, c and d, such that $ab + cd$ is a prime, say p.

Substitute $a = \dfrac{p - cd}{b}$ into the given equation, we have

$$p(p - 2cd + bc) = (b^2 + c^2)(b^2 + bd - d^2).$$

Since p is a prime, p divides $b^2 + c^2$ or $b^2 + bd - d^2$.

If $p \mid (b^2 + c^2)$, then

$$0 < b^2 + c^2 < 2ab < 2(ab + cd) = 2p$$

which yields $b^2 + c^2 = p$, namely,

$$ab + cd = b^2 + c^2.$$

Hence $b \mid c(c - d)$. Clearly, $\gcd(b, c) = 1$ (as $ab + cd$ is a prime), so $b \mid (c - d)$, contradicting $0 < c - d < c < b$.

If $p \mid (b^2 + bd - d^2)$, then $0 < b^2 + bd - d^2 < 2(ab + cd) = 2p$ which yields $b^2 + bd - d^2 = p$, that is, $ab + cd = b^2 + bd - d^2 = a^2 + ac - c^2$, so $a \mid (c + d)c$ and $b \mid (c + d)d$. But $\gcd(ab, cd) = 1$, so $c + d$ is divisible by both a and b. Since $0 < c + d < 2a$, and $0 < c + d < 2b$, thus $c + d = a$ and $c + d = b$, a contradiction.

3.5 Set $a - b = k$, then the given equation can be rewritten as

$$k(c - b) = b^2. \tag{A.2}$$

Set $\gcd(k, b - c) = d$. If $d > 1$, then d has a prime divisor p. From the above equation we have $p \mid b^2$, so $p \mid b$. Combining with $p \mid (b - c)$ and $p \mid k$ we get $p \mid c$ and $p \mid a$, contradicting $\gcd(a, b, c) = 1$. Hence $d = 1$ and (A.2) implies that k and $c - b$ are all perfect squares.

4.1 Let $x(x + 1)(x + 2)(x + 3) = y^2$, where x and y are positive integers. Then

$$(x^2 + 3x + 1)^2 - y^2 = 1,$$

and it is easy to check that it is impossible.

4.2 Assume that integer n can be expressed as the difference of two squares of integers: $n = x^2 - y^2$, that is, $n = (x - y)(x + y)$. Since $x + y$ and $x - y$ have the same parity, n is either odd or divisible by 4.

Conversely, if n is odd, we can take $x - y = 1$ and $x + y = n$, namely $x = \dfrac{n+1}{2}$ and $y = \dfrac{n-1}{2}$. If $4 \mid n$, we can take $x - y = 2$ and $x + y = \dfrac{n}{2}$, namely $x = \dfrac{n}{4} + 1$ and $y = \dfrac{n}{4} - 1$, then $x^2 - y^2 = n$.

4. 3   Eliminate x from the system of equations we have

$$8 - 9x - 9z + 3x^2 + 6xy + 3y^2 - x^2 y - xy^2 = 0,$$

and rewrite it as

$$8 - 3x(3 - x) - 3y(3 - x) + xy(3 - x) + y^2(3 - x) = 0,$$

that is, $(3 - x)(3x + 3y - xy - y^2) = 8$. So $(3 - x) \mid 8$, thus $3 - x = \pm 1,\ \pm 2,\ \pm 4,\ \pm 8$. Hence $x = -5,\ -1,\ 1,\ 2,\ 4,\ 5,\ 7,\ 11$.

Substitute them into the original equations and check them one by one, we can find all integer solutions $(x, y, z) = (1, 1, 1),\ (-5, 4, 4),\ (4, -5, 4),\ (4, 4, -5)$.

5. 1   By properties of combinatory numbers we have

$$\binom{m+n}{m} = \frac{m+n}{m}\binom{m+n-1}{m-1} = \frac{m+n}{m}\binom{m+n-1}{n},$$

so $m\dbinom{m+n}{m} = (m+n)\dbinom{m+n-1}{m-1}$. Since $\gcd(m, m+n) = \gcd(m, n) = 1$, $m \left| \dbinom{m+n-1}{n}\right.$.

5. 2   Let $n = x + (x+1) + \cdots + (x+k-1)$, where x is a positive integer and $k \geqslant 2$. That is,

$$(2x + k - 1)k = 2n. \tag{A. 3}$$

If n is a power of 2, then k and $2x - 1 + k$ are all powers of 2. But $2x - 1$ is odd, so $k = 1$, contradicting the given condition.

Conversely, if n is not a power of 2, say $n = 2^{m-1}(2t +1)$, $m \geqslant 1$, $t \geqslant 1$. When $t \geqslant 2^{m-1}$, we can take $k = 2^m$, $x = t + 1 - 2^{m-1}$. When $t < 2^{m-1}$, we can take $k = 2t + 1$, $x = 2^{m-1} - t$, then k and x are all positive integers and $k \geqslant 2$.

5.3 When n is even we can take $a = 2n$ and $b = n$. If n is odd, we assume that p is the least odd prime which does not divide n. Then $p - 1$ has either no odd prime divisors (i. e. , is a power of 2), or its odd prime divisors all divide n. Hence $a = pn$ and the number or different prime divisors of $b = (p - 1)n$ is equal to the number of different prime divisors of n plus 1.

5.4 Rewrite the equation in the question as

$$x! \cdot y! = z \cdot (z - 1)!.$$

Take $x = n$, $y = n! - 1$ and $z = n!$, then infinitely many integer solutions satisfying the condition follow.

5.5 We can construct it by induction. When $n = 2$ we can take $a_1 = 1$, $a_2 = 2$. Suppose that we have $a_1, \ldots, a_k$ satisfying the requirement when $n = k$, put b_0 to be the least common multiple of $a_1, \ldots, a_k, a_i - a_j (1 \leqslant i, j \leqslant k, i \neq j)$, then the $k + 1$ numbers

$$b_0, a_1 + b_0, \ldots, a_k + b_0$$

meet the requirement.

6.1 Denote by S the sum stated in the question. We replace $- 1$ at any vertex by $+1$, then there are four numbers in S, say a, b, c and d whose signs are changed. Denote by S' the sum of 14 changed numbers. Since $a + b + c + d \equiv 0 \pmod 2$, we have

$$S - S' = 2(a + b + c + d) \equiv 0 \pmod 4.$$

Repeat this process until all numbers at the vertices are $+ 1$. Thus $S \equiv 1 + 1 + \cdots + 1 = 14 \equiv 2 \pmod 4$, so $S \neq 0$.

6.2 Positive integers N consisting of $n - 1$ digits 1 and one digit 7 can be expressed as $N = A_n + 6 \times 10^k$, where $0 \leqslant k \leqslant n - 1$, A_n is the integer consisting of n digits 1.

When $3 | n$, the sum of digits in A_n is divisible by 3, so $3 | A_n$ and $3 | N$, but $N > 3$, thus in this case N is not a prime.

Now suppose $3 \nmid n$. Note that $10^6 \equiv 1 \pmod 7$, so we can classify n by modulo 6 to discuss the values of A_n modulo 7 (we do not need to consider the case $n \equiv 0, 3 \pmod 6$). It is easy to see, for $l \geqslant 0$,

$$A_{6l+1} = \frac{1}{9} \times (10^{6l+1} - 1) = \frac{1}{9} \times (10^{6l} - 1) \times 10 + \frac{1}{9} \times (10 - 1)$$

$$\equiv 1 \pmod 7,$$

$$A_{6l+2} \equiv 4, \ A_{6l+4} \equiv 5, \ A_{6l+5} \equiv 2 \pmod 7.$$

On the other hand, 10^0, 10^2, 10^4, 10^5 modulo 7 are congruent to 1, 2, 4, 5, respectively. Thus when $n > 6$ according to $n \equiv 1, 2, 4, 5 \pmod 6$, take $k = 0, 4, 5, 2$, respectively, we have

$$N = A_n + 6 \times 10^k \equiv A_n - 10^k \equiv 0 \pmod 7,$$

so N is not a prime. Hence n does not meet the requirement if $n > 5$. When $n \leqslant 5$, it is easy to check that only $n = 1, 2$ meet the requirement.

6.3 By $a^m \equiv 1 \pmod p$ we have $a^m = 1 + px$. Hence

$$a^{pm} = (1 + px)^p = 1 + p^2 x + \binom{p}{2} p^2 x^2 + \cdots \equiv 1 \pmod{p^2}.$$

$$\text{(A.4)}$$

Further, $a^{p-1} \equiv 1 \pmod{p^2}$, so $a^{(p-1)m} \equiv 1 \pmod{p^2}$, thus $a^{pm} \equiv a^m \pmod{p^2}$. Combining with (A.4) we know $a^m \equiv 1 \pmod{p^2}$.

6.4 We can assume $m > 1$. Denote by $\bar{x}_k$ the remainder of x_k divided by m. Consider the ordered pairs of the numbers

$$\langle \bar{x}_1, \bar{x}_2 \rangle, \ \langle \bar{x}_2, \bar{x}_3 \rangle, \ \ldots, \ \langle \bar{x}_n, \bar{x}_{n+1} \rangle. \qquad \text{(A.5)}$$

Since there are m^2 different ordered pairs of the remainders divided by m, there must be two pairs which are the same if we take the left most $m^2 + 1$ pairs in sequence (A.5). Suppose $\langle \bar{x}_i, \bar{x}_{i+1} \rangle$ is a pair which is equal to some other pair $\langle \bar{x}_j, \bar{x}_{j+1} \rangle$ and has the minimal index i ($j \leqslant m^2 + 1$), we have to prove that i must be 1. Otherwise, from

$$x_{i-1} = x_{i+1} - x_i, \ x_{j-1} = x_{j+1} - x_j$$

it follows that $x_{i-1} \equiv x_{j-1} \pmod m$, so $\langle \bar{x}_{i-1}, \bar{x}_i \rangle = \langle \bar{x}_{j-1}, \bar{x}_j \rangle$, contradicting the minimal property of i. Thus $i = 1$. Now from $\langle \bar{x}_j, \bar{x}_{j+1} \rangle = \langle \bar{x}_1, \bar{x}_2 \rangle = \langle 1, 1 \rangle$ it follows that $x_{j-1} \equiv x_{j+1} - x_j \equiv 1 - 1 \equiv 0 \pmod m$, namely, $m \mid x_{j-1} (1 < j - 1 \leqslant m^2)$.

7.1 By the condition we have

$$3(n - 1) = 4(2^{p-1} + 1)(2^{p-1} - 1). \tag{A.6}$$

As the prime $p > 3$, by the Fermat's little theorem we have $p \mid (2^{p-1} - 1)$. Combining with (A.6) we get $2p \mid (n - 1)$, thus $(2^{2p} - 1) \mid (2^{n-1} - 1)$. By the given condition we have $n \mid (2^{2p} - 1)$. Therefore $n \mid (2^{n-1} - 1)$.

7.2 Since there are infinitely many primes, we can choose $2n$ distinct primes $p_1, \ldots, p_{2n}$. Due to the Chinese Remainder Theorem, a system of congruences

$$x \equiv -k \pmod{p_{2k-1} p_{2k}}, \ k = 1, 2, \ldots, n,$$

has a positive integer solution. As for any k $(1 \leqslant k \leqslant n)$, $x + k$ has at least two distinct prime divisors, it is not a power of a prime.

7.3 Assume that $m = 11^i u$, $n = 11^j v$, where i, j are nonnegative integers, and u, v are positive integers not divisible by 11. We want to prove that $u = v$, thus $m = 11^{i-j} v$. If $u \neq v$, we can assume $u > v$. Since $\gcd(u, 11) = 1$, by the Chinese Remainder Theorem, there exists a positive integer x, such that

$$x \equiv 0 \pmod{u}, \ x \equiv -1 \pmod{11}, \tag{A.7}$$

that is, $x = 11k - 1$ (k is some positive integer). From (A.7) it follows that $\gcd(11k - 1, m) = \gcd(x, 11^i u) = u$, but $\gcd(11k - 1, n) = \gcd(x, 11^j v) \leqslant v < u$, contradicting the condition $\gcd(11k - 1, m) = \gcd(11k - 1, n)$, thus $u = v$.

8.1 We have to prove that any prime divisor p of F_k satisfies $p \equiv 1 \pmod{2^{k+1}}$. Clearly, $p \neq 2$. Suppose that the order of 2 modulo p is r, by $p \mid F_k$ we have

$$2^{2^k} \equiv -1 \pmod{p}, \tag{A.8}$$

so $2^{2^{k+1}} \equiv 1 \pmod{p}$, hence $r \mid 2^{k+1}$. Therefore r is a power of 2. Assume that $r = 2^l$, where $0 \leqslant l \leqslant k + 1$. If $l \leqslant k$, then from $2^{2^l} \equiv 1 \pmod{p}$ it follows that $2^{2^k} \equiv 1 \pmod{p}$. Combining with (A.8) we get $p = 2$, this is impossible. Hence $l = k + 1$. Furthermore, $2^{p-1} \equiv 1 \pmod{p}$, thus $r \mid (p - 1)$, and $2^{k+1} \mid (p - 1)$, namely, $p \equiv 1 \pmod{}$

2^{k+1}).

8.2 (1) Let r be the order of a modulo mn. By $a^r \equiv 1 \pmod{mn}$ we get $a^r \equiv 1 \pmod m$ and $a^r \equiv 1 \pmod n$. Thus $d_1 \mid r$ and $d_2 \mid r$, hence $[d_1, d_2] \mid r$. On the other hand, from $a^{d_1} \equiv 1 \pmod m$ and $a^{d_2} \equiv 1 \pmod n$ it follows that $a^{[d_1, d_2]} \equiv 1 \pmod m$ and $a^{[d_1, d_2]} \equiv 1 \pmod n$. Since $\gcd(m, n) = 1$, $a^{[d_1, d_2]} \equiv 1 \pmod{mn}$. Thus $r \mid [d_1, d_2]$. Combining these results we have $r = [d_1, d_2]$.

(2) By direct verification we obtain that the order of 3 modulo 2^4 is 4. Also, we have the order of 3 modulo 5 is 4, so by (1) of Example 5, we get that the order of 3 modulo 5^4 is 4×5^3. Hence from (1) of this exercise we know the order of 3 modulo 10^4 is $[4, 4 \times 5^3] = 500$.

8.3 We prove it by induction. When $k = 1, 2$, the result is clearly true. Assume that for $k \geqslant 3$ there exists n_0 such that $2^k \mid (3^{n_0} + 5)$, write $3^{n_0} = 2^k u - 5$. If u is even, then $2^{k+1} \mid (3^{n_0} + 5)$. In the following we assume that u is odd.

The key of the proof is to note that for $k \geqslant 3$ we have

$$3^{2^{k-2}} = 1 + 2^k v, \quad v \text{ is odd.}$$

(See (8.11) of Example 5 in Chapter 8.) Now we have

$$3^{n_0 + 2^{k-2}} = 3^{n_0} \cdot 3^{2^{k-2}} = (-5 + 2^k u)(1 + 2^k v)$$
$$= -5 + (u - 5v + 2^k uv) \cdot 2^k.$$

The number in the parentheses about is even, so 2^{k+1} divides $3^{n_0 + 2^{k-2}} + 5$. This completes the proof.

9.1 Rewrite the equation as

$$(2x + 3y)^2 = 17y^2 + 4 \times 122.$$

Apply modulo 17 and we have $(2x + 3y)^2 \equiv 12 \pmod{17}$. But it is easy to verify that a perfect square modulo 17 has only values 0, 1, 2, 4, 8, 9, 13, 15 and 16, not 12. Therefore, the original equation has no integer solutions.

9.2 Applying modulo 4 we know that the equation

$$12^m - 5^n = -7$$

has no positive integer solutions. Clearly, the equation

$$12^m - 5^n = 7 \tag{A.9}$$

has solution $m = n = 1$. In the following we prove that when $m > 1$ it has no positive integer solutions. (A.9) modulo 3 we have $-(-1)^n \equiv 1 \pmod 3$, so n is odd, and $5^n \equiv 5 \pmod 8$. Since $m \geqslant 2$, we have $8 \mid 12^m$. (A.9) modulo 8 we get $-5 \equiv 7 \pmod 8$, which is impossible. Therefore $m = 1$, thus $n = 1$.

9.3 When $p = 2$ or $p = 5$ the condition is not satisfied. Assume the prime $p > 2$ and $p \neq 5$. By the binomial theorem, we have

$$2^p + 3^p = 2^p + (5-2)^p = 5^p - \binom{p}{1} 5^{p-1} \times 2 + \cdots + 5 \binom{p}{p-1} 2^{p-1}$$
$$= 5^2 u + 5p \times 2^{p-1},$$

where u is an integer.

So $5 \parallel (2^p + 3^p)$, hence $2^p + 3^p$ is not a k-th power of an integer $(k > 1)$.

9.4 Obviously, the equation has solution $x = y = 1$. The equation modulo 4 we know that y must be odd. If $y > 1$, the equation modulo 9 we have

$$5^x \equiv 2 \pmod 9. \tag{A.10}$$

It is easy to find that for $x = 1, 2, \ldots,$ 5^x modulo 9 they are periodically $5, 7, 8, 4, 2, 1$. By (A.10) we know that x must have the form $6k + 5$. The equation modulo 7, it is easy to verify that for odd y we have

$$3^y \equiv 3, 5, 6 \pmod 7.$$

When $x = 6k + 5$, from the Fermat's little theorem it follows that $5^6 \equiv 1 \pmod 7$, so

$$5^x = 5^{6k+5} \equiv 5^5 \equiv 3 \pmod 7,$$

thus the two sides modulo 7 are not equal. Hence it has no solution when $y > 1$. Thus it has only one positive integer solution $y = 1$, $x = 1$.

10.1   Since $a \geqslant 3$, $a - 1$ has a prime divisor p. By Fermat's little theorem, we have $a^p \equiv a \equiv 1 \pmod{p}$. By induction, it is easy to prove that all $n = p^k (k = 1, 2, \ldots)$ meet the requirement. (We can compare the problem with Example 2 in Chapter 8.)

10.2   The given condition can be rewritten as

$$2^{n_2} \equiv 1 \pmod{n_1}, \ 2^{n_3} \equiv 1 \pmod{n_2}, \ \ldots, \ 2^{n_1} \equiv 1 \pmod{n_k}.$$

Let $D = [n_1, \ldots, n_k]$. From the above equation it follows that

$$2^D \equiv 1 \pmod{n_i}, \ i = 1, \ldots, k.$$

Hence, $2^D \equiv 1 \pmod{D}$. Thus by Example 2 in Chapter 8 we have $D = 1$. Therefore $n_1 = n_2 = \cdots = n_k = 1$.

10.3   Let $a^3 + b^3 = ma^2 b$, then $\left(\dfrac{a}{b}\right)^3 - m\left(\dfrac{a}{b}\right)^2 + 1 = 0$, namely, the rational number $\dfrac{a}{b}$ is a root of the following polynomial equation with integer coefficients and the leading coefficient being 1

$$x^3 - mx^2 + 1. \qquad (A.11)$$

So $\dfrac{a}{b}$ must be an integer. On the other hand, any integer root of equation (A.11) must divide the constant term 1, thus it is ± 1. Further, a, b are positive integers, so $\dfrac{a}{b} = 1$, that is, $a = b$.

10.4   Obviously, $y > 1$. The original equation can be factorized into

$$(y - 1)(y^n + y^{n-1} + \cdots + y + 1) = x^n. \qquad (A.12)$$

The key is to prove that $y - 1$ and $y^n + y^{n-1} + \cdots + y + 1$ are relatively prime. If their greatest common divisor $d > 1$, then d has a prime divisor p. By $y \equiv 1 \pmod{p}$ we know $y^i \equiv 1 \pmod{p}$. Hence we have

$$y^n + y^{n-1} + \cdots + y + 1 \equiv n + 1 \pmod{p},$$

thus $p \mid (n + 1)$. But from (A.12) it follows that $p \mid x^n$, thus prime $p \mid x$, contradicting $\gcd(x, n + 1) = 1$. Hence $d = 1$. Now it follows from (A.12) that there are positive integers a and b, such that

$$y - 1 = a^n, \; y^n + y^{n-1} + \cdots + y + 1 = b^n. \qquad (A.13)$$

But $y^n < y^n + y^{n-1} + \cdots + y + 1 < (y+1)^n$, that is, $y^n + y^{n-1} + \cdots + y + 1$ lies between two adjacent n-th powers. Thus it is not an n-th power of some integer, contradicting the proved result (A. 13).

10.5 Let $a = \dfrac{x}{z}$, $b = \dfrac{y}{z}$, x, y and z positive integers, and $\gcd(x, y, z) = 1$. Then the statement $a^n - b^n$ is an integer is equivalent to

$$x^n \equiv y^n \pmod{z^n}. \qquad (A.14)$$

We have to prove that $z = 1$, from this we know that a and b are integers.

Let $z > 1$, then z has prime divisors. If z has an odd prime divisor p, we denote by r the least positive integer such that $x^r \equiv y^r \pmod{p}$ holds. By (A. 14) we have $x^n \equiv y^n \pmod{p}$, so $r \mid n$ (see Remark 3 in Example 5 of Chapter 8). Let $p^\alpha \parallel n$ and $p^\beta \parallel (x^r - y^r)$ (note that as $a \neq b$, $x \neq y$). Then from (1) of Example 5 in Chapter 8 it follows that $p^{\alpha+\beta} \parallel (x^n - y^n)$, but (A. 14) implies that $p^n \mid (x^n - y^n)$, hence $p^n \leqslant p^{\alpha+\beta}$, so $n \leqslant \alpha + \beta$. Moreover, $p^\alpha \leqslant n$, so $\alpha \leqslant \log_p n$, thus

$$n \leqslant \log_p n + \beta,$$

which does not hold when n is sufficiently large (note that β is a fixed number). Therefore (A. 14) does not hold for infinitely many values of n, a contradiction.

If z has no odd prime divisor, then z is a power of 2. Combine with (A. 12) and $\gcd(x, y, z) = 1$ we know that x and y are odd numbers. When n is odd, from

$$x^n - y^n = (x - y)(x^{n-1} + x^{n-2}y + \cdots + xy^{n-2} + y^{n-1}),$$

noting that the second divisor of the right side of the above equation is odd, it follows that $2^n \mid (x^n - y^n)$ implies $2^n \mid (x - y)$ for $x \neq y$. Thus there are only finitely many such n. When n is even, let $2^s \parallel (x^2 - y^2)$, by (2) in Remark 3 of Example 5, Chapter 8, we know that if $2^\alpha \parallel n$, then $2^{\alpha+s-1} \parallel (x^n - y^n)$. Combining with (A. 12) we have

$n \leqslant \alpha + s - 1$. Since $\alpha \leqslant \log_2 n$, we have

$$n \leqslant \log_2 n + s - 1,$$

which does not hold for sufficiently large even values of n, a contradiction.

10. 6   The result is easy to prove if every a_i is not equal to n. Since $2n$ numbers

$$a_1, a_2, \ldots, a_n, 2n - a_1, 2n - a_2, \ldots, 2n - a_n$$

are all positive integers and less than $2n$, there must be two which are the same, namely, there are i and j such that $a_i = 2n - a_j$. Since $i = j$ implies that $a_i = n$, contradicting the assumption, we have $i \neq j$, thus $a_i + a_j = 2n$ is divisible by $2n$.

Now without loss of generality we assume $a_n = n$. Consider $n - 1$ ($\geqslant 3$) integers $a_1, a_2, \ldots, a_{n-1}$, among them there are two numbers whose difference is not divisible by n, since if all $\binom{n-1}{2}$ differences of these numbers are divisible by n, then $\binom{n-1}{2} \geqslant 3$ implies that there are three numbers $a_i < a_j < a_k$ such that $n \mid (a_j - a_i)$, $n \mid (a_k - a_j)$, thus $a_k - a_i = (a_k - a_j) + (a_j - a_i) \geqslant 2n$, which it impossible.

Without loss of generality we assume $a_1 - a - 2$ is not divisible by n. Consider the following n numbers

$$a_1, a_2, a_1 + a_2, a_1 + a_2 + a_3, \ldots, a_1 + a_2 + \cdots + a_{n-1}. \tag{A. 15}$$

If they are not congruent modulo n pairwise, then among them there is one number which is divisible by n. If there are two numbers in (A. 15) which are congruent modulo n, then the difference of these two numbers is divisible by n, which implies that there is a sum of some numbers of $a_1, \ldots, a_{n-1}$ divisible by n (as $a_1 - a_2$ is not divisible by n). Denote this sum by kn. If k is even, then the result holds. If k is odd, add a_n into the sum, we also get the result.

Printed in the USA
CPSIA information can be obtained
at www.ICGtesting.com
CBHW072118071224
18567CB00011B/355

9 789814 271141